MCDP 1-2

Campaigning

U.S. Marine Corps

PCN 142 000008 00

DEPARTMENT OF THE NAVY
Headquarters United States Marine Corps
Washington, D.C. 20380-1775

1 August 1997

FOREWORD

Tactical success in combat does not of itself guarantee victory
in war. *What matters ultimately in war is strategic success*:
attainment of our political aims and the protection of our national interests. The operational level of war provides the linkage between tactics and strategy. It is the discipline of
conceiving, focusing, and exploiting a variety of tactical actions to realize a strategic aim. With that thought as our point
of departure, this publication discusses the intermediate, *operational* level of war and the military campaign which is the
vehicle for organizing tactical actions to achieve strategic
objectives.

The Marine air-ground task force (MAGTF) clearly has operational as well as tactical capabilities. Thus it is essential
that Marine leaders learn to think operationally. Marine Corps
Doctrinal Publication (MCDP) 1-2, *Campaigning*, provides
the doctrinal basis for military campaigning in the Marine
Corps, particularly as it pertains to a Marine commander or a
MAGTF participating in the campaign. *Campaigning* applies

the warfighting philosophies in MCDP 1, *Warfighting*, specifi-
cally to the operational level of war. It is linked to the other
publications of the MCDP series and is fully compatible with
joint doctrine.

MCDP 1-2 supersedes Fleet Marine Force Manual (FMFM)
1-1, *Campaigning*, of 1990. MCDP 1-2 retains the spirit,
scope, and basic concepts of its predecessor. MCDP 1-2 further develops and refines some of those concepts based on re-
cent experiences, continued thinking about war, and the
evolving nature of campaigning in the post-Cold War world.

The new version of *Campaigning* has three significant additions: an expanded discussion of the linkage between strategic
objectives and the campaign, a section on conflict termination,
and a section titled "Synergy" that describes how key capabilities are harmonized in the conduct of a campaign to achieve the
strategic objective. These additions have been derived from the
development of the other doctrinal publications in the MCDP
series and joint doctrine.

Chapter 1 discusses the campaign and the operational level
of war, their relationship to strategy and tactics, and their relevance to the Marine Corps. Chapter 2 describes the process of
campaign design: deriving a military strategic aim from political objectives and constraints, developing a campaign concept
that supports our strategic objectives, and making a campaign
plan that translates the concept into a structured configuration
of actions required to carry out that concept. Chapter 3

discusses the actual conduct of a campaign and the problem of adapting our plans to events as they unfold.

Central to this publication is the idea that military action at any level must ultimately serve the demands of policy. Marine leaders at all levels must understand this point and must recognize that we pursue tactical success not for its own sake, but for the sake of larger political goals. Military strength is only one of several instruments of national power, all of which must be fully coordinated with one another in order to achieve our strategic and operational objectives. Marine leaders must be able to integrate military operations with the other instruments of national power.

This publication makes frequent use of historical examples. These examples are intended to illustrate teachings that have universal relevance and enduring applicability. No matter what the scope and nature of the next mission—general war or military operations other than war—the concepts and the thought processes described in this publication will apply. As with *Warfighting*, this publication is descriptive rather than prescriptive. Its concepts require judgment in applica- tion.

This publication is designed primarily for MAGTF commanders and their staffs and for officers serving on joint and combined staffs. However, commanders at all levels of any military organization require a broad perspective, an understanding of the interrelationships among the levels of war, and knowledge of the methods for devising and executing a progressive series of actions in pursuit of a distant objective in the face of hostile resistance. Marine officers of any grade and specialty can easily find themselves working—either directly or indirectly—for senior leaders with strategic or operational responsibilities. Those leaders need subordinates who understand their problems and their intentions. Therefore, as with MCDP 1, I expect all officers to read and reread this publication, understand its message, and apply it.

C. C. KRULAK
General, U.S. Marine Corps
Commandant of the Marine Corps

DISTRIBUTION: 142 000008 00

Throughout this publication, masculine nouns and pronouns are used for the sake of simplicity. Except where otherwise noted, these nouns and pronouns apply to either gender.

Campaigning

Chapter 1. The Campaign
Strategy—Tactics—Operations—Strategic-Operational
Connection—Tactical-Operational Connection—
Interdependence of the Levels of War—Campaigns—
Battles and Engagements—A Comparative Case
Study: Grant Versus Lee–*Policy–Military Strategy–
Operations in 1864–Tactics*—The Marine Corps and
Campaigning

Chapter 2. Designing the Campaign
Supporting the Military Strategic Aim–*Campaigning
Under an Annihilation Strategy–Campaigning Under
an Erosion Strategy*—Identifying the Enemy's Critical
Vulnerabilities—The Campaign Concept–*Phasing the
Campaign–Conceptual, Functional, and Detailed
Planning–Conflict Termination*—Campaign Design:
Two Examples–*Case Study: The Recapture of Europe,
1944-45–Case Study: Malaysia, 1948-60*—The Campaign
Plan

Chapter 1

The Campaign

"Battles have been stated by some writers to be the chief and deciding features of war. This assertion is not strictly true, as armies have been destroyed by strategic operations without the occurrence of pitched battles, by a succession of inconsiderable affairs."[1]

—Henri Jomini

"For even if a decisive battle be the goal, the aim of strategy must be to bring about this battle under the most advantageous circumstances. And the more advantageous the circumstances, the less, proportionately, will be the fighting."[2]

—B. H. Liddell Hart

"It is essential to relate what is strategically desirable to what is tactically possible with the forces at your disposal. To this end it is necessary to decide the development of operations before the initial blow is delivered."[3]

—Bernard Montgomery

T his book is about military campaigning. A *campaign* is a series of related military operations aimed at accomplishing a strategic or operational objective within a given time and space.[4] A *campaign plan* describes how time, space, and purpose connect these operations.[5] Usually, a campaign is aimed at achieving some particular strategic result within a specific geographic theater. A war or other sustained conflict sometimes consists of a single campaign, sometimes of several. If there is more than one campaign, these can run either in sequence or—if there is more than one theater of war—simultaneously. Campaigning reflects the operational level of war, where the results of individual tactical actions are combined to fulfill the needs of strategy.

Military campaigns are not conducted in a vacuum. Military power is employed in conjunction with other instruments of national power—diplomatic, economic, and informational—to achieve strategic objectives. Depending upon the nature of the operation, the military campaign may be the main effort, or it may be used to support diplomatic or economic efforts. The military campaign must be coordinated with the nonmilitary efforts to ensure that all actions work in harmony to achieve the ends of policy. Frequently, particularly in military operations other than war, the military campaign is so closely integrated with other government operations that these nonmilitary actions can be considered to be part of the campaign.

In this chapter, we will describe how events at different levels of war are interrelated, focusing on the operational level as the link between strategy and tactics. We will examine the campaign as the basic tool of commanders at the operational level and discuss its relevance to the Marine Corps.

STRATEGY

War grows out of political conflict. Political policy determines the aims of each combatant's strategy and directs each side's conduct. Thus, as Liddell Hart wrote, "any study of the problem ought to begin and end with the question of policy."[6] Strategy is the result of intellectual activity that strives to win the objectives of policy by action in peace as in war.

National strategy is the art and science of developing and using the political, economic, and informational powers of a nation, together with its armed forces, during peace and war, to secure national objectives. National strategy connotes a global perspective, but it requires coordination of all the elements of national power at the regional or theater level as well. Because a campaign takes place within a designated geographic theater and may involve nonmilitary as well as military elements, campaign design is often equivalent to theater strategy.

4

Military strategy is the art and science of employing the armed forces of a nation to secure the objectives of national policy by the application of force or the threat of force. It involves the establishment of military strategic objectives, the allocation of resources, the imposition of conditions on the use of force, and the development of war plans.[7]

Strategy is both a product and a process. That is, strategy involves both the creation of plans—specific strategies to deal with specific problems—and the process of implementing them in a dynamic, changing environment. Therefore, strategy requires both detailed planning and energetic adaptation to evolving events.

Strategic concepts describe the ways in which the elements of national power are to be used in the accomplishment of our strategic ends, i.e., our policy objectives.[8] U.S. military strategy is implemented by the combatant commanders and is always joint in nature. In practice, the execution of our military strategy in any particular region requires coordination—and often considerable compromise—with other governmental agencies, with allies, with members of coalitions formed to meet specific contingencies, and with nongovernmental organizations.

Military strategy must be subordinate to national strategy and must be coordinated with the use of the nonmilitary instruments of our national power. Historically, we have sometimes found it difficult to maintain those relationships correctly, and we have sometimes fought in the absence of a clear national or military strategy.

TACTICS

Marines are generally most familiar—and therefore most comfortable—with the tactical domain, which is concerned with defeating an enemy force through fighting at a specific time and place.[9] The tactical level of war is the province of combat. The means of tactics are the various elements of combat power at our disposal. Its ways are the concepts by which we apply that combat power against our adversary. These concepts are sometimes themselves called tactics—in our case, tactics founded on maneuver. The goal of tactics is victory: defeating the enemy force opposing us. In this respect, we can view tactics as the discipline of winning battles and engagements.

The tactical level of war includes the maneuver of forces in contact with the enemy to gain a fighting advantage, the application and coordination of fires, the sustainment of forces throughout combat, the immediate exploitation of success to seal the victory, the combination of different arms and weapons, the gathering and dissemination of pertinent information, and the technical application of combat power within a tactical action—all to cause the enemy's defeat.

In practice, the events of combat form a continuous fabric of activity. Nonetheless, each tactical action, large or small, can generally be seen as a distinct episode fought within a distinct space and over a particular span of time.

Tactical success does not of itself guarantee success in war. In modern times, victory in a single battle is seldom sufficient to achieve strategic victory as it sometimes was in Napoleon's time. In fact, a single battle can rarely determine the outcome of a campaign, much less that of an entire war. Even a succession of tactical victories does not necessarily ensure strategic victory, the obvious example being the American military experience in Vietnam. Accordingly, we must recognize that *defeating the enemy in combat cannot be viewed as an end in itself, but rather must be considered merely a means to a larger end.*

OPERATIONS

It follows from our discussions of the strategic and tactical levels of war that there is a level of the military art above and distinct from the realm of tactics and subordinate to the domain of strategy. This level is called the *operational level of war.* It is the link between strategy and tactics.[10] Action at the operational level aims to give meaning to tactical actions in the context of some larger design that is itself framed by strategy. Put

7

another way, *our aim at the operational level is to get strate-gically meaningful results from tactical efforts.*

Thus at the operational level of war we conceive, focus, and exploit a variety of tactical actions in order to attain a strategic goal. In its essence, the operational level involves deciding when, where, for what purposes, and under what conditions to give battle—or to refuse battle—in order to fulfill the strategic goal. Operations govern the deployment of forces, their com-mitment to or withdrawal from combat, and the sequencing of successive tactical actions to achieve strategic objectives.

The nature of these tasks requires that the operational com-mander retain a certain amount of latitude in the conception and execution of plans. "The basic concept of a campaign plan should be born in the mind of the man who has to direct that campaign."[11] If higher authority overly prescribes the concept of operations, then the commander becomes a mere executor of tactical tasks instead of the link between those tasks and the strategic objectives. Such was the case in many U.S. air opera-tions over North Vietnam.

The term "operations" implies broader dimensions of time and space than does "tactics" because a strategic orientation forces the operational commander to consider a perspective broader than the limits of immediate combat.[12] While the tacti-cian fights the battle, the operational commander must look be-yond the battle—seeking to shape events in advance in order to create the most favorable conditions possible for future combat actions. The operational commander likewise seeks to take maximum advantage of the outcome of any actual combat (win, lose, or draw), finding ways to exploit the resulting situation to the greatest strategic advantage.

Although the operational level of war is sometimes described as large-unit tactics, it is erroneous to define the operational level according to echelon of command. Military actions need not be of large scale or involve extensive combat to have an important political impact.[13] The distance between tactical actions and their strategic effects varies greatly from conflict to conflict. In World War II, for example, strategic effects could usually be obtained only from the operations of whole armies or fleets. In a future very large-scale conventional conflict, a corps commander may well be the lowest-level operational commander. In Somalia, on the other hand, strategic (i.e., political) effects could result from the actions of squads or even individuals. *Regardless of the size of a military force or the scope of a tactical action, if it is being used to directly achieve a strategic objective, then it is being employed at the operational level.*

STRATEGIC-OPERATIONAL CONNECTION

No level of war is self-contained. Strategic, operational, and tactical commanders, forces, and events are continually interacting with one another. Although we may view the chain of command as a hierarchical pyramid in which directives and power flow from higher to lower, in fact the command structure is often more like a spider web: a tug at any point may have an impact throughout the structure. Information must therefore flow freely in all directions. To use a different metaphor, the fingers have to know what the brain is feeling for, and the brain has to know what the fingers are actually touching.

We must always remember that the political end state envisioned by policy makers determines the strategic goals of all military actions. We must also understand that the relationship between strategy and operations runs both ways. That is, just as strategy shapes the design of the campaign, so must strategy adapt to operational circumstances and events.

Strategy guides operations in three basic ways: it establishes aims, allocates resources, and imposes restraints and constraints on military action. Together with the nature and actions of the enemy and the characteristics of the area of operations, strategic guidance defines the parameters within which we can conduct operations.

First, strategy translates policy objectives into military terms by establishing the military strategic aim. What political effect must our military forces achieve? What enemy assets must our tactical forces seize, neutralize, threaten, or actually destroy in order to either bend the enemy to our will or break him completely? The operational commander's principal task is to determine and pursue the sequence of actions that will most directly accomplish the military strategic mission. It is important to keep in mind that the military strategic aim is but one part of a broader national strategy.

Strategists must be prepared to modify aims in the light of actual developments, as they reevaluate costs, capabilities, and expectations. While required to pursue the established aim, the operational commander is obliged to communicate the associated risks to superiors. When aims are unclear, the commander must seek clarification and convey the impact— positive or negative—of continued ambiguity.

Second, strategy provides resources, both tangible resources such as material and personnel and intangible resources such as political and public support for military operations. When resources are insufficient despite all that skill, talent, dedication, and creativity can do, the operational commander must seek additional resources or request modification of the aims.

Third, strategy, because it is influenced by political and social concerns, places conditions on the conduct of military operations. These conditions take the form of restraints and constraints. Restraints prohibit or restrict certain military actions such as the prohibition imposed on MacArthur against bombing targets north of the Yalu River in Korea in 1950 or the United States' policy not to make first use of chemical weapons in World War II. Restraints may be constant, as the laws of warfare, or situational, as rules of engagement. Constraints, on the other hand, obligate the commander to certain military courses of action such as President Jefferson Davis's decision that the policy of the Confederacy would be to hold as much territory as possible rather than employ a more flexible defense or resort to wide-scale guerrilla tactics, or the decision that the Arab members of the Coalition should be the liberators of Kuwait City during the Gulf War. Similarly, strategy may constrain the commander to operations which gain rapid victory such as Abraham Lincoln's perceived need to end the American Civil War quickly lest Northern popular resolve falter.

When conditions imposed by strategy are so severe as to prevent the attainment of the established aim, the commander must request relaxation of either the aims or the limitations. However, we should not be automatically critical of conditions imposed on operations by higher authority, since "policy is the guiding intelligence"[14] for the use of military force. Nonetheless, no senior commander can use the conditions imposed by higher authority as an excuse for military failure.

TACTICAL-OPERATIONAL CONNECTION

Just as strategy shapes the design of the campaign while simultaneously adapting to operational circumstances and events, so operations must interact with tactics. Operational plans and directives that are rooted in political and strategic aims establish the necessary focus and goals for tactical actions. Operational planning provides the context for tactical decisionmaking. Without this operational coherence, warfare at the tactical level is reduced to a series of disconnected and unfocused tactical actions. Just as operations must serve strategy by combining tactical actions so as to most effectively and economically achieve the aim, they must also serve tactics by creating the most advantageous conditions for our tactical actions. In other words, we try to shape the situation so that the outcome is merely a matter of course. "Therefore," Sun Tzu said, "a skilled commander seeks victory from the situation and does not demand it of his subordinates."[15] Just as we must continually interface with strategy to gain our direction, we must also maintain the flexibility to adapt to tactical circumstances as they develop, for tactical results will impact on the conduct of the campaign. As the campaign forms the framework for combat, so do tactical results shape the conduct of the campaign. In this regard, the task is to exploit tactical developments—victories, draws, even defeats—to strategic advantage.

INTERDEPENDENCE OF THE LEVELS OF WAR

The levels of war form a hierarchy. Tactical engagements are components of battle, and battles are elements of a campaign. The campaign, in turn, is itself but one phase of a strategic design for gaining the objectives of policy. While a clear hierarchy exists, there are no sharp boundaries between the levels. Rather, they merge together and form a continuum.

Consequently, a particular echelon of command is not necessarily concerned with only one level of war. A theater commander's concerns are clearly both strategic and operational. A Marine air-ground task force commander's responsibilities will be operational in some situations and largely tactical in others and may actually span the transition from tactics to operations in still others. A commander's responsibilities within the hierarchy depend on the scale and nature of the conflict and may shift up and down as the war develops.

Actions at one level can often influence the situation at other levels.[16] Harmony among the various levels tends to reinforce success, while disharmony tends to negate success. Obviously, failure at one level tends naturally to lessen success at the other levels.

It is perhaps less obvious that the tactics employed to win in actual combat may prevent success at a higher level. Imagine a government whose strategy is to quell a growing insurgency by isolating the insurgents from the population but whose military tactics cause extensive collateral death and damage. The government's tactics alienate the population and make the enemy's cause more appealing, strengthening him politically and therefore strategically.

Brilliance at one level of war may to some extent overcome shortcomings at another, but rarely can it overcome outright incompetence. Operational competence is meaningless without the ability to achieve results at the tactical level. Strategic incompetence can squander what operational success has gained.

The natural flow of influence in the hierarchy is greatest at the top. That is, it is much more likely that strategic incompetence will squander operational and tactical success than that tactical and operational brilliance will overcome strategic incompetence or disadvantage. The Germans are widely considered to have been tactically and operationally superior in the two World Wars. Their strategic incompetence, however, proved an insurmountable obstacle to victory. Conversely, outgunned and overmatched tactically, the Vietnamese Communists prevailed strategically.

The flow can work in reverse as well: brilliance at one level can overcome, at least in part, shortcomings at a higher level. In this way, during the American Civil War, the tacti- cal and operational abilities of Confederate military leaders in the eastern theater of war held off the strategic advantages of the North for a time until President Lincoln found a commander—General Grant—who would press those advantages. Similarly, in North Africa, early in World War II, the tactical and operational flair of German General Erwin Rommel's Africa Corps negated Britain's strategic advantage only for a time.

What matters finally is success at the strategic level. The concerns of policy are the motives for war in the first place, and it is the political impact of our operations that determines our success or failure in war. It is far less important to be able to discern at what level a certain activity takes place or where the transition between levels occurs than to ensure that from top to bottom and bottom to top all the components of our military effort are in harmony. We must never view the tactical domain in isolation because the results of combat become relevant only in the larger context of the campaign. The campaign, in turn, gains meaning only in the context of strategy.

CAMPAIGNS

The principal tool by which the operational commander pursues the conditions that will achieve the strategic goal is the *campaign*. Campaigns tend to take place over the course of weeks or months, but they may span years. They may vary drastically in scale from large campaigns conceived and controlled at the theater or even National Command Authorities level to smaller campaigns conducted by joint task forces within a combatant command. Separate campaigns may be waged sequentially within the same conflict, each pursuing intermediate objectives on the way to the final strategic goal. It is also possible to pursue several campaigns simultaneously if there are multiple theaters of war. In modern times, for each

U.S. conflict or military operation other than war there is normally only one campaign at a time within one geographic theater of war or theater of operations.[17] That campaign is always
joint in character and falls under the command of either a regional commander in chief or a subordinate joint force commander. The joint force commander's campaign is made up of
a series of related major operations, some of which may be
conducted by a single Service.

In the past, however, the word "campaign" has been used
very flexibly. Historians often refer to lesser campaigns within
larger ones. For example, the Allied Pacific campaign in the
Second World War comprised subordinate campaigns by General Douglas MacArthur in the Southwest Pacific, Admiral
William Halsey in the South Pacific, and Admiral Chester
Nimitz in the Central Pacific. Halsey's campaign in the South
Pacific itself included a smaller campaign in the Solomon Islands that lasted 5 months and consisted of operations from
Guadalcanal to Bougainville. Similarly, we often hear of "air
operations" or "submarine operations" as if they constituted independent campaigns. Nonetheless, while the Desert Storm
campaign had an initial phase dominated by aerial forces, we
do not refer to this as an air campaign.

At times, the relationships of these operations may not be
readily apparent. They may seem to be isolated tactical events
such as Operation Eldorado Canyon, the punative U.S. air-
strike against Libya in 1986. On the surface, this operation ap-
peared to be a single military response to a specific Libyan act,

the bombing of the La Belle discotheque in Berlin in which two U.S. servicemen were killed and a number injured. In fact, this operation was part of much larger series of actions intended to attain the strategic objective of reducing or eliminating Libya's sponsorship of international terrorism. Nonmilitary actions included efforts to isolate Libya diplomatically coupled with economic sanctions and information to publicize Libya's support of terrorism. Military actions consisted of a series of freedom of navigation operations conducted in the Gulf of Sidra that showed U.S. military commitment and put more pressure on the Libyan government.[18]

BATTLES AND ENGAGEMENTS

A *battle* is a series of related tactical engagements. Battles last longer than engagements and involve larger forces. They occur when adversaries commit to fight to a decision at a particular time and place for a significant objective. Conse- quently, battles are usually operationally significant (though not necessarily operationally decisive).[19] This is not always so. The Battle of the Somme in 1916, which was actually a series of inconclusive battles over the span of 4½ months, merely moved the front some 8 miles while inflicting approximately 1 million casualties on the opposing armies.

An *engagement* is a small tactical conflict, usually between maneuver forces.[20] Several engagements may compose a battle.

Engagements may or may not be operationally significant, although our intent is to gain advantage from the results.

Battles and engagements are the armed collisions that mark potential turning points in a campaign. While such combat provides perceptible structure, it is the campaign design that gives combat meaning. In some campaigns, military forces play a supporting role and are not really the main effort, as in the campaign to isolate Iraq following the Gulf War. In that case, tactical actions are small, infrequent, and undertaken largely to enforce political and economic sanctions and to maintain blockades. Even in campaigns where military forces represent the main effort, sometimes small engagements are so continuous and large battles so rare that a campaign may seem to be one drawn-out combat action. For instance, we often refer to the Allies' World War II campaign against German submarines in the Atlantic as the "Battle of the Atlantic." Guerrilla wars and insurgencies often follow a similar pattern. The structure of campaigns in such cases is sometimes hard to perceive because the ebb and flow in the antagonists' fortunes happen bit by bit rather than in sudden, dramatic events.

Even when a campaign involves distinct battles, operational and strategic advantage can be gained despite tactical defeat. General Nathaniel Greene's campaign against the British in the Carolinas during the American Revolution provides an example. In the winter of 1781, Greene maneuvered his army for almost 2 months to avoid engagement with the British force commanded by Lord Cornwallis. In March of 1781, reinforced

19

by Continental soldiers, militia, and riflemen from Virginia and North Carolina, Greene decided to challenge the British in North Carolina at Guilford Courthouse. The Americans fought well, inflicting more casualties than they sustained, but were forced to withdraw from the field. This engagement, a defeat for Greene, proved to be a turning point in the campaign.[21] The British, exhausted from the previous pursuit and short on supplies, were unable to exploit their tactical victory and withdrew to the coast, leaving their scattered South Carolina garrisons vulnerable.[22]

The point is that victory in battle is only one possible means to a larger end. The object should be to accomplish the aim of strategy with as little combat as practicable, reducing "fighting to the slenderest possible proportions."[23]

However, none of this is to say that we can—or should try to—avoid fighting on general principle. How much fighting we do varies according to the strength, skill, intentions, and determination of the opposing sides. The ideal is to give battle only where we want and when we must—when we are at an advantage and have something important to gain that we cannot gain without fighting. However, since we are opposed by a hostile will with ideas of his own, we do not always have this option. Sometimes we must fight at a disadvantage when forced to by a skilled enemy or when political obligations constrain us (as would have been the case had the North Atlantic Treaty Organization's plan for the forward defense of Germany against the old Warsaw Pact been executed).

A COMPARATIVE CASE STUDY: GRANT VERSUS LEE

A comparative examination of the strategic, operational, and tactical approaches of Generals Ulysses S. Grant and Robert E. Lee during the American Civil War offers an interesting illustration of the interaction of the levels. Popular history regards Grant as a butcher and Lee a military genius. A study of their understanding of the needs of policy and the consistency of their strategic, operational, and tactical methods casts the issue in a different light.[24]

Policy

The North faced a demanding and complex political problem, namely "to reassert its authority over a vast territorial empire, far too extensive to be completely occupied or thoroughly controlled."[25] Furthermore, President Lincoln recognized that Northern popular resolve might be limited and established rapid victory as a condition as well. Lincoln's original policy of conciliation having failed, the President opted for the unconditional surrender of the South as the only acceptable aim. His search for a general who would devise a strategy to attain his aim ended with Grant in March 1864. By comparison, the South's policy aim was to preserve its newly declared independence. The South's strategic aim was simply to prevent the North from succeeding, to make the endeavor more costly than the North was willing to bear.

Military Strategy

The South's policy objectives would seem to dictate a military strategy of erosion aimed at prolonging the war as a means to breaking Northern resolve. In fact, this was the strategy preferred by Confederate President Jefferson Davis. Such a strategy would require close coordination of the Southern armies and a careful husbanding of the Confederacy's inferior resources. In practice, however, no Southern general in chief was appointed until Lee's appointment in early 1865. No doubt it was in part because of the Confederacy's basic political philosophy of states' rights that the military resources of the various Southern states were poorly distributed. Campaigns in the various theaters of war were conducted almost independently.

Lee's decision to concentrate his army in northern Virginia reflected a perspective much narrower than Grant's and the fact that he was politically constrained to defend Richmond. However, this decision was due also to Lee's insistence on an offensive strategy—not merely an offensive defense as in the early stages of the war but eventually an ambitious offensive strategy in 1862 and '63 aimed at invading the North as a means to breaking Northern will. (See figure.) Given the South's relative weakness, Lee's strategy was questionable at best[26]—both as a viable means of attaining the South's policy aims and also in regard to operational practicability, particularly the South's logistical ability to sustain offensive campaigns.

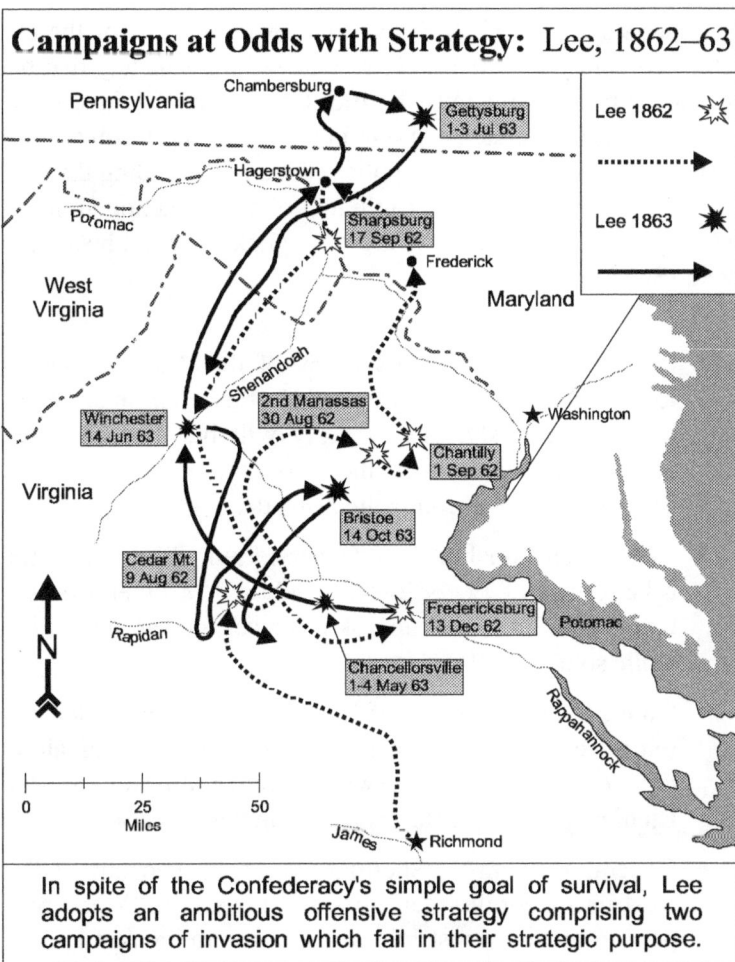

Campaigns at Odds with Strategy: Lee, 1862–63

Pennsylvania

Chambersburg

Gettysburg
1-3 Jul 63

Lee 1862

Hagerstown

Potomac

Sharpsburg
17 Sep 62

Lee 1863

Frederick

West
Virginia

Maryland

Shenandoah

2nd Manassas
30 Aug 62

Washington

Winchester
14 Jun 63

Chantilly
1 Sep 62

Virginia

Bristoe
14 Oct 63

Cedar Mt.
9 Aug 62

Fredericksburg
13 Dec 62

Potomac

N

Rapidan

Chancellorsville
1-4 May 63

Rappahannock

0   25   50
Miles

James

Richmond

In spite of the Confederacy's simple goal of survival, Lee adopts an ambitious offensive strategy comprising two campaigns of invasion which fail in their strategic purpose.

Grant's strategy of 1864 was directly supportive of the established policy objectives. He recognized immediately that his
military strategic aim must be the destruction of Lee's army,
and he devised a strategy of annihilation focused resolutely on
that aim. Consistent with the policy objective of ending the war
as rapidly as possible, Grant initiated offensive action simultaneously on all fronts to close the ring quickly around his opponent. (See figure.)

- General George Meade's Army of the Potomac was to
  lock horns with Lee's Army of Northern Virginia, pursuing it relentlessly. "Lee's army will be your objective
  point. Wherever he goes, there you will go also."[27] Gran-
  t's headquarters accompanied Meade.

- In the Shenandoah Valley, General Franz Sigel was to fix
  a large part of Lee's forces in place. "In other words,"
  Grant said, "if Sigel can't skin himself he can hold a leg
  while some one else skins."[28]

- On the Peninsula, south of Richmond, General Butler was
  reinforced by troops taken from occupation duties along
  the Southern coast. He was to move up and threaten
  Richmond from a different direction than Meade.

- General William T. Sherman was to sweep out of the
  west into Georgia, then up along the coast. "You I pro-
  pose to move against Johnston's army, to break it up and
  to get into the interior of the enemy's country as far as
  you can, inflicting all the damage you can against their

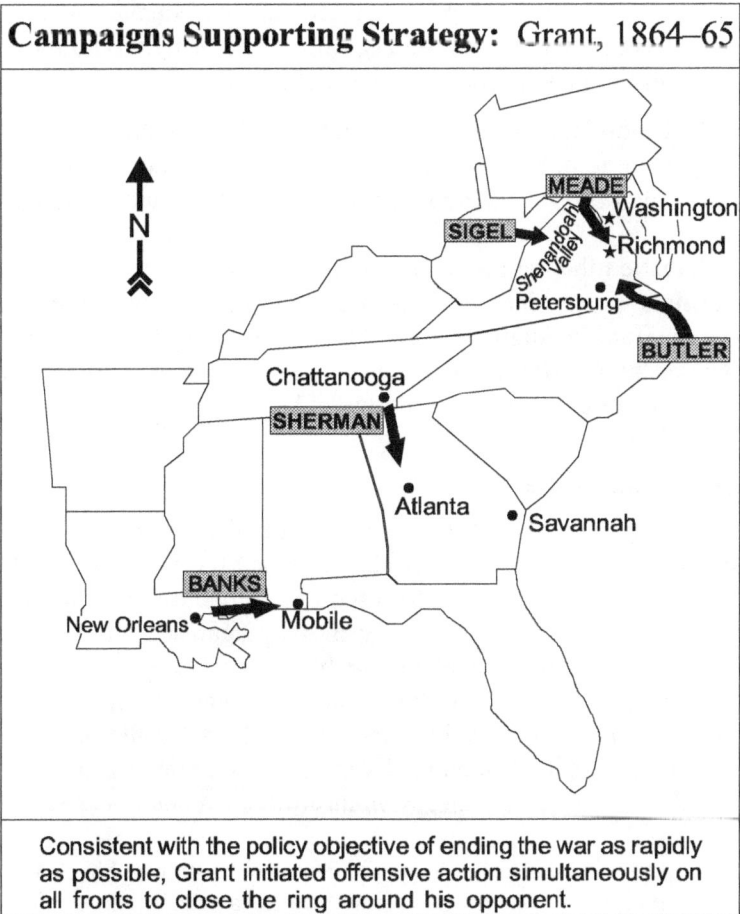

Campaigns Supporting Strategy: Grant, 1864–65

Consistent with the policy objective of ending the war as rapidly as possible, Grant initiated offensive action simultaneously on all fronts to close the ring around his opponent.

war resources."[29] After eliminating Confederate forces in Georgia and the Carolinas, Sherman's army would move north in a strategic envelopment of Lee.

- Union land and sea forces in the vicinity of New Orleans were to concentrate and take Mobile, Alabama, thus cutting off one of the last functioning Confederate seaports.

Satisfied that he had finally found a commander who could translate policy into a successful military strategy, Lincoln wrote Grant in August 1864: " 'The particulars of your plans I neither know nor seek to know. . . . I wish not to obtrude any restraints or constraints upon you.' "[30]

Operations in 1864

Consistent with his strategy of grinding Lee down as quickly as possible and recognizing his ability to pay the numerical cost, Grant aggressively sought to force Lee frequently into pitched battle. He accomplished this by moving against Richmond in such a way as to compel Lee to block him. Grant never fell back to lick his wounds but rather continued relentlessly to press his fundamental advantages no matter what the outcome of a particular engagement. Even so, it is unfair to discount Grant, as some have done, as an unskilled butcher:

He showed himself free from the common fixation of his contemporaries upon the Napoleonic battle as the hinge upon which warfare must turn. Instead, he developed a highly uncommon ability to rise above the fortunes of a single battle and to master the flow of a long series of events, almost to the

26

point of making any outcome of a single battle, victory, draw, or even defeat, serve his eventual purpose equally well.[31]

Lee, on the other hand, had stated that having the weaker force, his desire was to avoid a general engagement.[32] In practice, however, he seemed unable to resist the temptation of a climactic Napoleonic battle whenever the enemy was within reach. Despite a number of tactical successes, Lee was eventually pinned to the fortifications at Petersburg, where he was besieged by Grant from mid-June 1864 to early April 1865. Lee's eventual attempt to escape from Petersburg led to his army's capture at Appomattox on 9 April 1865. (See figure on page 28.)

The most important subordinate campaign, other than that of the Army of the Potomac itself, was Sherman's. His initial opponent, General Joseph Johnston, in contrast to Lee, seemed to appreciate the Confederacy's need to protract the conflict. Johnston—

fought a war of defensive maneuver, seeking opportunities to fall upon enemy detachments which might expose themselves and inviting the enemy to provide him with such openings, meanwhile moving from one strong defensive position to another in order to invite the enemy to squander his resources in frontal attacks, but never remaining stationary long enough to risk being outflanked or entrapped.[33]

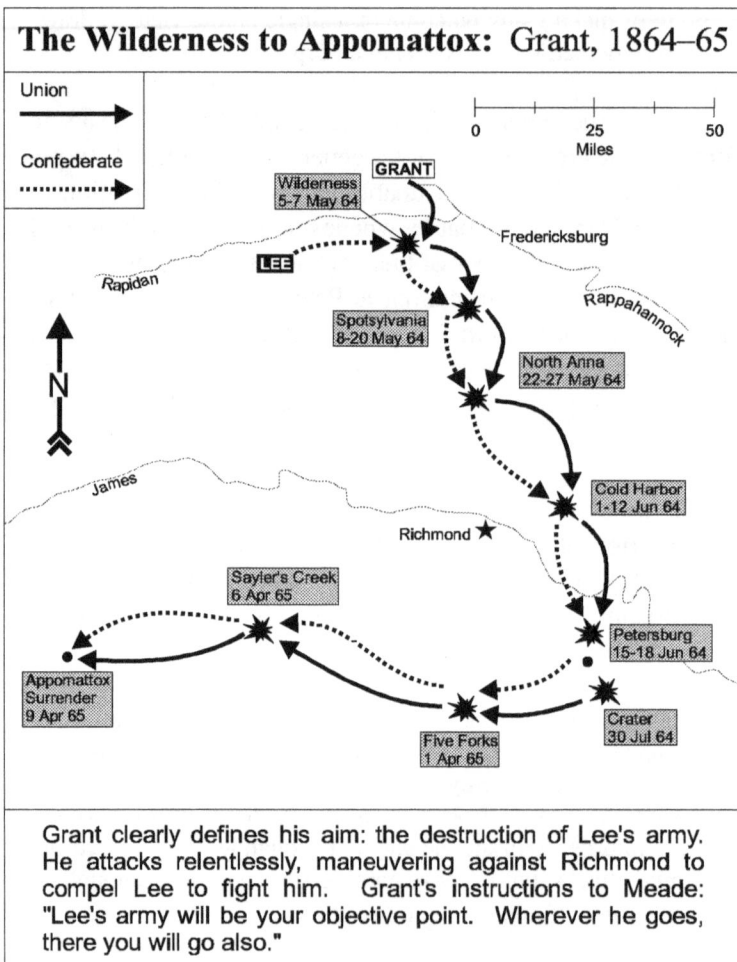

The Wilderness to Appomattox: Grant, 1864–65

Grant clearly defines his aim: the destruction of Lee's army. He attacks relentlessly, maneuvering against Richmond to compel Lee to fight him. Grant's instructions to Meade: "Lee's army will be your objective point. Wherever he goes, there you will go also."

Between Chattanooga and Atlanta, while suffering minimal casualties, Johnston held Sherman to an average advance of a mile a day. Of Johnston's campaign, Grant himself wrote—

> For my own part, I think that Johnston's tactics were right. Anything that could have prolonged the war a year beyond the time that it did finally close, would probably have exhausted the North to such an extent that they might have abandoned the contest and agreed to a separation.[34]

Tactics

Lee's dramatic tactical successes in battles such as Second Manassas and Chancellorsville speak for themselves. Nevertheless, neither Lee nor Grant can be described as particularly innovative tactically. In fact, both were largely ignorant of the technical impact of the rifled bore on the close-order tactics of the day, and both suffered high casualties as a result.[35] However, due to the relative strategic situations, Grant could better absorb the losses that resulted from this tactical ignorance than could Lee, whose army was being bled to death. In this way, Grant's strategic advantage carried down to the tactical level.

Grant's activities at all levels seem to have been mutually supporting and focused on the objectives of policy. Lee's strategy and operations appear to have been, at least in part, incompatible with each other, with the requirements of policy, and with the realities of combat. In the final analysis, Lee's tactical flair could not overcome operational and strategic shortcomings of the Confederacy.

THE MARINE CORPS AND CAMPAIGNING

Having described how goals at the different levels of war interact and introduced the campaign, we must now ask ourselves what is the relevance of this subject to the Marine Corps. We can answer this question from several perspectives. Marine airground task forces (MAGTFs) will participate in campaigns, and Marines will serve on joint staffs and participate in the design of campaigns. MAGTF commanders and their staffs may find themselves designing major operations in support of a campaign.

Organizationally, the MAGTF is uniquely equipped to perform a variety of tactical actions—amphibious, air, and land—and to sequence or combine those actions in a coherent scheme. The MAGTF's organic aviation allows the commander to project power in depth and to shape events in time and space. The command structure with separate headquarters for the tactical control of ground, air, and logistics actions frees the MAGTF command element to focus on the operational conduct of war.

A MAGTF is often the first American force to arrive in an undeveloped theater of operations. In that case, the MAGTF commander will often have operational-level responsibilities regardless of the size of the MAGTF. In some cases, the MAGTF may provide the nucleus of a joint task force headquarters. Even in a developed theater, a MAGTF may be required to conduct major operations as part of a larger campaign in pursuit of a strategic objective. The commander of a MAGTF must be prepared to describe its most effective operational employment in a joint or multinational campaign.

The news media, because of its global reach and ability to influence popular opinion, can have operational effects—that is, it can often elevate even minor tactical acts to political importance. Consequently, Marines must understand how tactical action impacts on politics; this is the essence of understanding war at the operational level.

Finally, regardless of the echelon of command or scale of activity, even if an action rests firmly in the tactical realm, the methodology described here—devising and executing a progressive series of actions in pursuit of a goal and deciding when and where to fight for that goal—applies.

Chapter 2

Designing the Campaign

"By looking on each engagement as part of a series, at least insofar as events are predictable, the commander is always on the high road to his goal."[1]

—Carl von Clausewitz

"To be practical, any plan must take account of the enemy's power to frustrate it; the best chance of overcoming such obstruction is to have a plan that can be easily varied to fit the circumstances met; to keep such adaptability, while still keeping the initiative, the best way is to operate along a line which offers alternative objectives."[2]

—B. H. Liddell Hart

H aving defined and described the operational level of war and the campaign, we will now discuss the mental process and the most important considerations required to plan a campaign. The commander's key responsibility is to provide focus. Through the campaign plan, the commander fuses a variety of disparate forces and tactical actions, extended over time and space, into a single, coherent whole.[3]

SUPPORTING THE MILITARY STRATEGIC AIM

Campaign design begins with the military strategic aim. The campaign design should focus all the various efforts of the campaign on the established strategic aim. Effective campaign planners understand the role of the campaign under consideration in the context of the larger conflict. They also understand the need to resolve, to the extent possible, any ambiguities in the role of our military forces. This focus on the military strategic aim is the single most important element of campaign design.

There are only two ways to use military force to impose our political will on an enemy.[4] The first approach is to make the enemy helpless to resist the imposition of our will through the destruction of his military capabilities. Our aim is the elimination (permanent or temporary) of the enemy's military capacity—which does not necessarily mean the physical destruction of all his forces. We call this a military *strategy of annihilation.*[5] We use force in this way when we seek an *un-limited political objective*—that is, when we seek to overthrow the enemy leadership or force its unconditional surrender. We may also use it in pursuit of a more *limited political objective* if we believe that the enemy will continue to resist our demands as long as he has any means to do so.

The second approach is to convince the enemy that making peace on our terms will be less painful than continuing to fight. We call this a *strategy of erosion*—the use of our military means with the aim of wearing down the enemy leadership's will to continue the struggle.[6] In such a strategy, we use our military forces to raise the enemy's costs higher than he is will-ing to pay. We use force in this manner in pursuit of limited political goals that we believe the enemy leadership will ulti-mately be willing to accept. (See figure.)

All military strategies fall into one of these fundamental categories. Campaign planners must understand the chosen strategy and its implications at the operational level. Failure to understand the basic strategic approach (annihilation or ero-sion) will prevent the development of a coherent campaign plan and may cause military and diplomatic leaders to work at cross-purposes.

Campaigning Under an Annihilation Strategy
If the policy aim is to destroy the enemy's political entity—to overthrow his political structure and impose a new one—then

our military aim *must* be annihilation.[7] Even if our political goal is more limited, however, we may still seek to eliminate the enemy's capacity to resist. In the Gulf War, we completely destroyed the ability of Iraqi forces to resist us in the Kuwaiti theater of operations, but we did not overthrow the enemy regime. Our political goal of liberating Kuwait was limited, but our military objective, *in the Kuwaiti theater of operations*, was not.[8] In the Falklands war, Britain had no need to attack

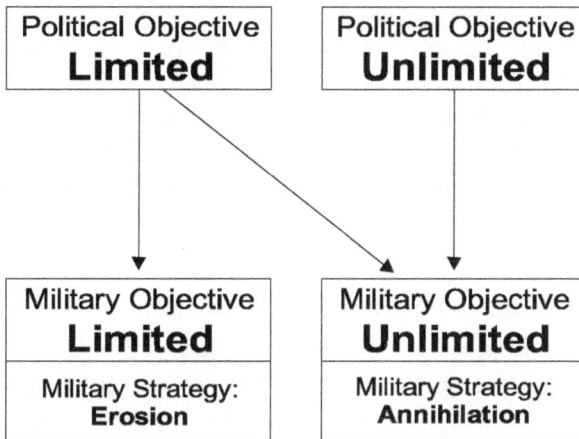

Determining Military Strategy.

the Argentine mainland or to overthrow its government in order
to recover the disputed islands. In the area of operations, however, the British isolated and annihilated the Argentine forces.

Strategies of annihilation have the virtue of conceptual simplicity. The focus of our operational efforts is the enemy armed
forces. Our intent is to render them powerless. We may choose
to annihilate those forces through battle or through destruction
of the infrastructures that support them. Our main effort re-
sides in our own armed forces. The other instruments of national   power—diplomatic,   economic,   and
informational—clearly support it.[9] Victory is easily measured:
when the enemy's fighting forces are no longer able to present
organized resistance, we have achieved military victory. Regardless of whether our political goal is limited or unlimited, a
strategy of annihilation puts us into a position to impose our
will.

Campaigning Under an Erosion Strategy

Erosion strategies are appropriate when our political goal is
limited and does not require the destruction of the enemy leadership, government, or state. Successful examples of erosion
appear in the American strategy against Britain during the
American Revolution and the Vietnamese Communist strategy
against France and the United States in Indochina.

Erosion strategies involve a great many more variables than annihilation strategies. These distinctions are important and critical to the campaign planner. In erosion strategies, we have a much wider choice in our operational main efforts, the relationship of military force to the other instruments of power is much more variable, and our definition of victory is much more flexible.

The means by which a campaign of erosion convinces the enemy leadership to negotiate is the infliction of unacceptable costs. Note that we mean unacceptable costs to the leadership, not to the enemy population. Our actions must have an impact on the enemy leadership. We must ask ourselves:

- What does the enemy leadership value?

- How can we threaten it in ways the enemy leadership will take seriously?

Often, the most attractive objective for a campaign of erosion is the enemy's military forces. Many regimes depend on their military forces for protection against their neighbors or their own people. If we substantially weaken those forces, we leave the enemy leadership vulnerable.

In erosion strategies, however, we may choose a nonmilitary focus for our efforts. Instead of threatening the enemy leadership's survival by weakening them militarily, we may seize or neutralize some other asset they value—and prove that we can maintain our control. Our objective may be a piece of territory that has economic, political, cultural, or prestige value; shipping; trade in general; financial assets; and so on. The aim to seize and hold territory normally makes our military forces the main effort. Successful embargoes and the freezing of financial assets, on the other hand, depend primarily on diplomacy and economic power. In the latter examples, therefore, military forces play a supporting role and may not be engaged in active combat operations at all.

We may also seek to undermine the leadership's prestige or credibility. Special forces and other unconventional military elements may play a role in such a campaign, but the main effort will be based on the informational and diplomatic instruments of our national power.

Victory in a campaign of erosion can be more flexibly defined and/or more ambiguous than is victory in a campaign of annihilation. The enemy's submission to our demands may be explicit or implicit, embodied in a formal treaty or in behind-the-scenes agreements. If we are convinced that we have made our point, changed his mind or his goals, or have so eroded the enemy's power that he can no longer threaten us, we may simply "declare victory and go home." Such conclusions may seem unsatisfying to military professionals, but they are acceptable if they meet the needs of national policy.

IDENTIFYING THE ENEMY'S CRITICAL VULNERABILITIES

Economy demands that we focus our efforts toward some object or factor of decisive importance in order to achieve the greatest effect at the least cost. Differing strategic goals may dictate different kinds of operational targets. If we are pursuing an erosion strategy, we will seek objectives that raise to unacceptable levels the cost to the enemy leadership of noncompliance with our demands. Depending on the nature of the enemy leadership, our objectives may be the military forces or their supporting infrastructure, the internal security apparatus, territorial holdings, economic assets, or something else of value to our specific enemy. If we are pursuing a strategy of annihilation, we will seek objectives that will lead to the collapse of his military capabilities.

In either case, we must understand both the sources of the enemy's strength and the key points at which he is vulnerable. We call a key source of strength a *center of gravity*. It represents something without which the enemy cannot function.[10]

We must distinguish between a *strategic center of gravity* and an *operational center of gravity*. The former is an objective whose seizure, destruction, or neutralization will have a profound impact on the enemy leadership's will or ability to continue the struggle. Clausewitz put it this way—

For Alexander, Gustavus Adolphus, Charles XII, and Frederick the Great, the center of gravity was their army. If the army had been destroyed, they would all have gone down in history as failures.[11] In countries subject to domestic strife, the center of gravity is generally the capital. In small countries that rely on large ones, it is usually the army of their protector. Among alliances, it lies in the community of interest, and in popular uprisings [the centers of gravity are] the personalities of the leaders and public opinion. It is against these that our energies should be directed.[12]

An operational center of gravity, on the other hand, is normally an element of the enemy's armed forces. It is that concentration of the enemy's military power that is most dangerous to us or the one that stands between us and the accomplishment of our strategic mission. The degree of danger a force poses may depend on its size or particular capabilities, its location relative to ourselves, or the particular skill or enterprise of its leader.[13]

The strategic and operational centers of gravity may be one and the same thing, or they may be very distinct. For example, think of the campaign of 1864 in the case study in chapter 1. Sherman's strategic objectives were the destruction of the South's warmaking resources and will to continue the war. Until Johnston's Army of Tennessee was disposed of, Sherman's army had to stay concentrated and could not disperse over a wide enough area to seriously affect the South's economic infrastructure. For Sherman, Johnston's Army represented the operational, but not the strategic, center of gravity.

Usually we do not wish to attack an enemy's strengths directly because that exposes us to his power. Rather, we seek to attack his weaknesses in a way that avoids his strength and minimizes the risk to ourselves. Therefore we seek some *critical vulnerability*. A critical vulnerability is related to, but not the same as, a center of gravity; the concepts are complementary. A vulnerability cannot be *critical* unless it undermines a key strength. It also must be something that we are capable of attacking effectively.

Critical vulnerabilities may not be immediately open to attack. We may have to create vulnerability—to design a progressive sequence of actions to expose or isolate it, creating over time an opportunity to strike the decisive blow. An example would be to peel away the enemy's air defenses in order to permit a successful attack on his key command and control facilities.

Just as we ruthlessly pursue our enemy's critical vulnerabilities, we should expect him to attack ours. We must take steps to protect or reduce our vulnerabilities over the course of the campaign. This focus on the enemy's critical vulnerabilities is central to campaign design.

In order to identify the enemy's center of gravity and critical vulnerabilities, we must have a thorough understanding of the enemy. Obtaining this understanding is not simple or easy. Two of the most difficult things to do in war are to develop a realistic understanding of the enemy's true character and capabilities and to take into account the way that our forces and actions appear from his viewpoint. Instead, we tend to turn him into a stereotype—a cardboard cut-out or strawman—or, conversely, to imagine him 10-feet tall. We often ascribe to him attitudes and reflexes that are either mirror images of our own or simply fantasies—what we would like him to be or to do, rather than what his own particular situation and character would imply that he is. This insufficient thought and imagination makes it very difficult to develop realistic enemy courses of action, effective deception plans or ruses, or high-probability branches and sequels to our plans. In designing our campaign, we must understand the unique characteristics of our enemy and focus our planning to exploit weaknesses derived from that understanding.

THE CAMPAIGN CONCEPT

After determining whether the strategic aim is erosion or annihilation, describing its application in the situation at hand, and identifying the enemy's centers of gravity and critical vulnerabilities that we will attack to most economically effect the enemy's submission or collapse, we must now develop a campaign concept. This concept captures the essence of our design and provides the foundation for the campaign plan. It expresses in clear, concise, conceptual language a broad vision of what we plan to accomplish and how we plan to do it. Our intent, clearly and explicitly stated, is an integral component of the concept. Our concept should also contain in general terms an idea of when, where, and under what conditions we intend to give or refuse battle.

The concept should demonstrate a certain boldness, for boldness is in itself "a genuinely creative force."[14] It should focus on the enemy's critical vulnerabilities. It should exhibit creativity and avoid discernible conventions and patterns; make use of artifice, ambiguity, and deception; and reflect, as Churchill wrote, "an original and sinister touch, which leaves the enemy puzzled as well as beaten."[15] It should create multiple options so that we can adjust to changing events and so that the enemy cannot discern our true intentions. It should be as simple as the situation allows. It should provide for speed in execution—which is a weapon in itself.

Each campaign should have a single, unifying concept. Often a simple but superior idea has provided the basis for success. Grant's plan of fixing Lee near Richmond while loosing Sherman through the heart of the South was one such idea. The idea of bypassing Japanese strongholds in the Pacific became the basis for the Americans' island-hopping campaigns in the Second World War. MacArthur's bold, simple concept of a seaborne, operational turning movement became the Inchon landing in 1950.

Phasing the Campaign

A campaign is required whenever we pursue a strategic aim not attainable through a single tactical action at a single place and time. A campaign therefore includes several related phases that

45

may be executed simultaneously or in sequence. A campaign may also have several aspects, each to be executed by different forces or different *kinds* of forces. Phases are a way of organizing the diverse, extended, and dispersed activities of the campaign. As Eisenhower pointed out, "These phases of a plan do not comprise rigid instructions, they are merely guideposts. . . . Rigidity inevitably defeats itself, and the analysts who point to a changed detail as evidence of a plan's weakness are completely unaware of the characteristics of the battlefield."[16]

Each phase may constitute a single operation or a series of operations. Our task is to devise a combination of actions that most effectively and quickly achieve the strategic aim. While each phase may be distinguishable from the others as an identifiable episode, each is necessarily linked to the others and gains significance only in the larger context of the campaign. The manner of distinction may be separation in time or space or a difference in aim or in forces assigned.

We should view each phase as an essential component in a connected string of events that are related in cause and effect. Like a chess player, we must learn to think beyond the next move, to look ahead several moves, and to consider the long-term effects of those moves and how to exploit them. We cannot move without considering the enemy's reactions or anticipations, unlikely as well as likely.

Because each phase involves one or more decision points, we must think through as far as practicable the possible branches or options resulting from each decision. Such decision

points are often represented by battles, which—despite everything we can do to predetermine their outcome—can be either
lost or won. Each branch from a decision point will require different actions on our part and each action demands various
follow-ups—sequels or potential sequels.[17] "The higher commander must constantly plan, as each operation progresses, so
to direct his formations that success finds his troops in proper
position and condition to undertake successive steps without
pause."[18]

Each phase of the campaign is aimed at some intermediate
goal necessary to the accomplishment of the larger aim of the
campaign. Each phase has its own distinct intent which con-
tributes to the overall intent of the campaign. Generally speak-
ing, the phasing of a campaign should be event-driven rather
than schedule-driven. Each phase should represent a natural
subdivision of the campaign; we should not break the campaign
down into numerous arbitrary chunks that can lead to a plod-
ding, incremental approach sacrificing tempo.

The process of developing a sequence of phases in a campaign operates in two directions simultaneously: forward and
backward.[19] We begin our planning with both the current situation and the desired end state in mind—recognizing, of course,
that the end state may change as the situation unfolds. We plan
ahead, envisioning mutually supporting phases, whose combined effects set the stage for the eventual decisive action. At
the same time, however, and as a check on our planning, we
envision a reasonable set of phases backward from the end

state toward the present. The two sets of phases, forward and backward, have to mesh.

Phasing, whether sequential or simultaneous, allows us to allocate resources effectively over time. Taking the long view, we must ensure that resources will be available when needed in the later stages of the campaign. Effective phasing must take into account the process of logistical culmination. If resources are insufficient to sustain the force until the accomplishment of the strategic aim, logistical considerations may demand that the campaign be organized into sequential phases. Each of these must be supportable in turn, each phase followed by a logistical resupply or buildup. Moreover, logistical requirements may dictate the purpose of certain phases as well as the sequence of those phases.

Resource availability depends in large part on time schedules—such as sustainment or deployment rates—rather than on the events of war. Therefore, as we develop our intended phases, we must reconcile the time-oriented phasing of resource availability with the event-driven phasing of operations.

Conceptual, Functional, and Detailed Planning

The process of creating a broad scheme for accomplishing our goal is called *conceptual planning*. To translate the campaign concept into a complete and practicable plan requires functional planning and detailed planning. *Functional planning*, as the name implies, is concerned with designing the functional components necessary to support the concept: the subordinate concepts for command and control, maneuver, fires, intelligence, logistics, and force protection.[20] Functional planning ensures that we work through the feasibility of the campaign concept with respect to every functional area.

Detailed planning encompasses the specific planning activities necessary to ensure that the plan is coordinated: specific command relationships, movements, landing tables, deployment or resupply schedules, communications plans, reconnaissance plans, control measures, etc. Detailed design should not become so specific, however, that it inhibits flexi- bility.

No amount of subsequent planning can reduce the requirement for an overall concept. While conceptual planning is the foundation for functional and detailed planning, the process works in the other direction as well. Our concept must be adaptable to functional realities. Functional planning in turn must be sensitive to details of execution. The operational concept (a conceptual concern) should be used to develop the deployment plan (a functional concern). However, the realities of deployment schedules sometimes dictate employment schemes. Campaign design thus becomes a two-way process aimed at harmonizing the various levels of design activity.

The farther ahead we project, the less certain and detailed should be our design. We may plan the initial phase of a campaign with some degree of certainty, developing extensive functional and detailed plans. However, since the results of that phase will shape the phases that follow, subsequent plans must be increasingly general. The plan for future phases will be largely conceptual, perhaps consisting of no more than a general intent and several contingencies and options.

Conflict Termination

Two of the most important aspects of campaign design are defining the desired end state and planning a transition to post-conflict operations. Every campaign and every strategic effort have a goal. Every military action eventually ends.

The decisions when and under what circumstances to suspend or terminate combat operations are, of course, political decisions. Military leaders, however, are participants in the decisionmaking process. It is their responsibility to ensure that political leaders understand both the existing situation and the implications—immediate and long-term, military *and* political—of a suspension of combat at any point in the conflict. In 1864, for example, Union commanders understood well that any armistice for the purposes of North-South negotiation would likely mark an end to Union hopes. Regardless of the theoretical gap between the military and the political realms, combat operations, once halted, would have been virtually impossible to restart.[21] In the 1991 Gulf War, the timing of conflict termination reflected the achievement of our political and military aims in the Kuwaiti theater of op- erations.

Campaign designers must plan for conflict termination from the earliest possible moment and update these plans as the campaign evolves. What constitutes an acceptable political and military end state, the achievement of which will justify a termination to our combat operations? In examining any proposed end state, we must consider whether it guarantees an end to the fundamental problems that brought on the struggle in the first place, or whether instead it leaves in place the seeds of further conflict. If the latter, we must ask whether the chosen method of termination permits our unilateral resumption of military operations. Most practical resolutions of any conflict involve some degree of risk. Military leaders must always be prepared to ask the political leadership whether the political benefits of an early peace settlement outweigh the military risks—and thus also the political risks—of accepting a less-than-ideal conclusion to hostilities.

When addressing conflict termination, commanders must consider a wide variety of operational issues including disengagement, force protection, transition to postconflict operations, and reconstitution and redeployment. Thorough campaign planning can reduce the chaos and confusion inherent in abruptly ending combat operations. When we disengage and end combat operations, it is of paramount importance to provide for the security of our forces as well as noncombatants and enemy forces under our control. The violent emotions of war cannot be quelled instantly, and various friendly and enemy forces may attempt to continue hostile actions.

Once combat operations cease, the focus will likely shift to military operations other than war. The scope of these operations ranges from peacekeeping and refugee control to mine clearing and ordnance disposal to food distribution. Repairing host nation infrastructure and restoring host country control are operational-level concerns. Commanders at all levels must coordinate their efforts with a variety of governmental, non-governmental, and host nation agencies.

A final issue to be addressed in conflict termination is reconstitution and redeployment. Reconstitution begins in theater. Units are brought to a state of readiness commensurate with the mission requirements and available resources. The results of combat will dictate whether this is done through the shifting of internal resources within a degraded unit (reorganization) or the rebuilding of a unit through large-scale replacements (regeneration).[22] The capability to reconstitute and redeploy is especially important to naval expeditionary forces who must be able to complete one mission, reembark, and move on to the next task without hesitation. Regardless of the methods, reconstitution and redeployment pose a complex and demanding leadership and logistics challenge.

CAMPAIGN DESIGN: TWO EXAMPLES

The design of each campaign is unique. The campaign design is shaped first and foremost by the overall national strategy and the military strategic aim. The nature of the enemy, the characteristics of the theater of operations, and the resources available all influence the exact nature of each design. Nevertheless, the basic concepts of campaign design apply in any situation. Consider the following two case studies. While the designs of these two campaigns are radically different, the end result is the same: successful attainment of the strategic aim.

Case Study: The Recapture of Europe, 1944–45

An excellent example of campaign design during a major conflict is Eisenhower's broad plan for the recapture of Europe in the Second World War. The strategy was one of annihilation with the aim of eliminating Germany's military capacity. The design focused on the German forces as the primary center of gravity, although it recognized the importance of both political and economic centers such as Berlin and the Ruhr. The design employed a series of phases that were carried out in sequence as the campaign gained momentum and progressed towards the accomplishment of the ultimate objective. Eisenhower described this campaign design as "successive moves with possible alternatives."[23] (See figure, page 54.)

Land on the Normandy coast.

Build up the resources needed for a decisive battle in the Normandy-Brittany region and break out of the enemy's encircling positions. . . .

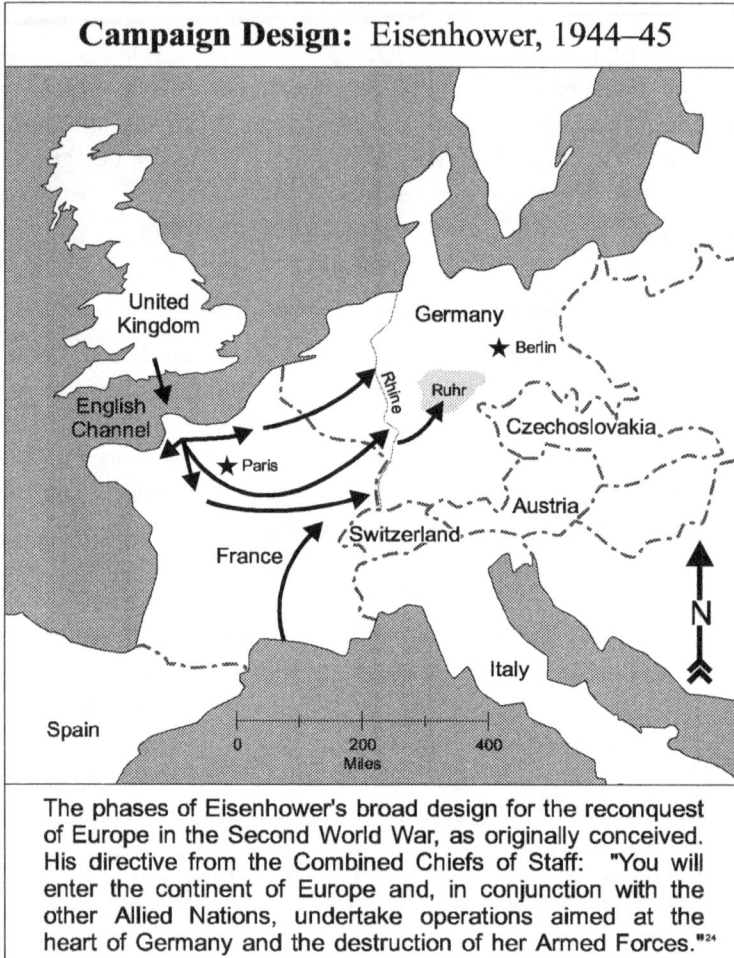

Campaign Design: Eisenhower, 1944–45

The phases of Eisenhower's broad design for the reconquest of Europe in the Second World War, as originally conceived. His directive from the Combined Chiefs of Staff: "You will enter the continent of Europe and, in conjunction with the other Allied Nations, undertake operations aimed at the heart of Germany and the destruction of her Armed Forces."[24]

Pursue on a broad front with two army groups, emphasizing the left to gain necessary ports and reach the boundaries of Germany and threaten the Ruhr. On our right we would link up with the forces that were to invade France from the south.

Build up our new base along the western border of Germany, by securing ports in Belgium and in Brittany as well as in the Mediterranean.

While building up our forces for the final battles, keep up an unrelenting offensive to the extent of our means, both to wear down the enemy and to gain advantages for the final fighting.

Complete the destruction of enemy forces west of the Rhine, in the meantime constantly seeking bridgeheads across the river.

Launch the final attack as a double envelopment of the Ruhr, again emphasizing the left, and follow this up by an immediate thrust through Germany, with the specific direction to be determined at the time.

Clean out the remainder of Germany.[25]

Eisenhower remarked that "this general plan, carefully outlined at staff meetings before D-Day, was never abandoned, even momentarily, throughout the campaign."[26]

Case Study: Malaysia, 1948–60

An example of campaign design very different from Eisenhower's can be found in the British campaign against a Communist insurgency in Malaysia. This example demonstrates that the concepts used to design a campaign in conventional conflicts apply as well in military operations other than war. While the British strategy was also one of annihilation, the nature of the conflict and the characteristics of the enemy dictated that the strategy had to be carried out over a much longer period in order to be successful. The centers of gravity and critical vulnerabilities were not primarily military in nature. Since this campaign was conducted over a number of years, the phases or building blocks of the campaign had to be pursued simultaneously rather than sequentially.

Both sides had clear goals and a clear concept for the political and military phasing of the struggle. The British had promised Malaysia its independence. Their goal was to leave a stable, non-Communist government in place after their departure. The Communists' goal was to obtain such a powerful military and political position within Malaysia that the British withdrawal would leave them dominant in the country. The British identified the center of gravity of the Communist movement as the large, impoverished Chinese minority who furnished the vast bulk of recruits for both the political and military wings of the Communist Party. Overall, the movement's critical vulnerability was its ethnic isolation in the Malay-dominated country. Militarily, its critical vulnerability was the dependence of Communist military units on food and other supplies from the widely scattered Chinese farming population. The center of gravity of the British-backed Malaysian government, on the other hand, was its claim to legitimacy and its promise of a better life than Communism could offer.

The British launched a multipronged campaign against the Communists. The navy insured that external support did not reach the Communists by sea. The army was responsible for keeping organized enemy units in the jungle, away from the population base and food supplies of the settled agricultural areas. The Malaysian government forces recognized that the jungle gave the enemy strength: enemy bases were hard to find and easily relocated if discovered. Search-and-destroy efforts were counterproductive because British strike forces were easily detected as they thrashed through the bush. This permitted the enemy not only to escape but to lay ambushes. However, the enemy's forces needed to move through the jungle as well, especially to obtain food. This made their forces vulnerable. The British knew where the food was grown and the routes the enemy supply columns had to follow to obtain it. Accordingly, the government forces themselves came to concentrate—very successfully—on the tactic of ambush.

Meanwhile, the police forces (recruited from the Malaysian population to a much greater size than the army) concentrated on providing security in the populated areas. They did this under very strict rules of engagement respecting the rights of the citizens, thus upholding the legality and legitimacy of constituted authority. Simultaneously, the destitute Chinese population was concentrated in clean, secure, well-designed new settlements, provided with the economic means to build homes in their own style, and given legal title to those homes and to adequate farmlands. This resettlement policy cut the guerrilla forces off from sources of recruits and, perhaps more important, food. The resettlement effort was accompanied by a political program to ensure that the Chinese minority obtained rights of citizenship equal to those of the Malay majority.

In combination, these patient and thoroughly coordinated military, police, economic, and political operations isolated the Communists both physically and psychologically from the main population. Despite some tactical successes (which included killing the first British commander in an ambush), the Malaysian Communist military forces were annihilated and the Party eliminated as a factor in Malaysian politics.[27]

Despite the obvious differences in the designs of these two campaigns, they both applied the basic concepts of campaign design to achieve the desired strategic objective. While the type of conflict and the nature of the enemy were radically different, both campaign designs had a clearly identified strategic aim, both focused on the enemy's centers of gravity and critical vulnerabilities, and both employed a campaign concept with appropriate phases tailored to accomplish the strategic aim.

THE CAMPAIGN PLAN

The campaign plan is the statement of the design for prosecuting the commander's portion of the overall strategy. It flows directly from the campaign concept and translates the concept into a structured configuration of actions required to carry out that concept. The plan describes a sequence of related operations that lead to a well-defined military end state. The campaign plan is a mechanism that provides focus and direction to subordinates.[28]

The campaign plan must be built around the strategy. It should describe, to subordinates and seniors alike, the end state which will attain the strategic aim. It must present the overall intent and concept of the campaign; a tentative sequence of phases and operational objectives which will lead to success; and general concepts for key supporting functions, especially a logistical concept that will sustain the force throughout the campaign. The logistical concept is vital since logistics, perhaps more than any other functional concern, can dictate what is operationally feasible.

The plan may describe the initial phases of the campaign with some certainty. However, the design for succeeding phases will become increasingly general as uncertainty grows and the situation becomes increasingly unpredictable. We must build as much adaptability as we can into the design of the campaign plan. Nevertheless, the final phase, the anticipated decisive action which will achieve final success and toward which the entire campaign builds, should be clearly envisioned and described. The campaign plan should establish tentative milestones and provide a measure of progress. It is not, however, a schedule in any final, immutable sense. Until the final aim is realized, we must continually adapt our campaign plan to changing interim aims (ours and the enemy's), results, resources, and limiting factors.

Above all, the campaign plan should be concise. General MacArthur's plan for his Southwest Pacific theater of operations was only four pages.[29] The campaign plan does not describe the execution of phases in tactical detail. Rather, it provides a framework for developing operation orders that in turn provide the tactical details.

Chapter 3

Conducting the Campaign

"A prince or general can best demonstrate his genius by managing a campaign exactly to suit his objectives and his resources, doing neither too much nor too little."[1]

—Carl von Clausewitz

"We must make this campaign an exceedingly active one. Only thus can a weaker country cope with a stronger; it must make up in activity what it lacks in strength."[2]

—Stonewall Jackson

B ecause campaign design is continuous, there is no point at which campaign design ceases and campaign execution begins. In fact, design and conduct are interdependent. Just as our design shapes our execution, so do the results of execution cause us to modify our design even in the midst of execution. Only with this thought firmly in mind can we proceed to discuss campaign execution.

Reduced to its essence, the art of campaigning consists of deciding who, when, and where to fight and for what purpose. Equally important, it involves deciding who, when, and where not to fight. It is, as Clausewitz described, "the use of engagements for the object of the war."[3]

STRATEGIC ORIENTATION

The conduct of politics and diplomacy continues in all its complexity even when military operations are under way. Sometimes the political situation is simple, and military operations can proceed in a straightforward fashion. It is increasingly common, however, for commanders even at the tactical level to find themselves navigating on terrain as complex politically as it is physically—cluttered with a confusing array of enemies, allies, neutrals, nongovernmental organizations, private volunteer organizations, United Nations forces and observers, and the press.

The art of campaigning means understanding when military force is our main effort and when it is acting in support of some other instrument of our national power. Thus, in the conduct as well as the design of a campaign, the overriding consideration is an unwavering focus on the goals of our strategy. The aims, resources, and conditions established by strategy are the filter through which we must view all our actions. Joint force commanders who may function anywhere from the theater to the tactical level must make their operational and tactical decisions with the theater strategy in mind. Lower-echelon commanders must understand the strategic context of their tactical missions if they are to provide useful feedback to higher levels on the effectiveness of field operations. Consequently, our strategic goals must be communicated clearly to commanders at every level.

THE USE OF COMBAT

Because tactical success alone does not guarantee the attainment of strategic goals, there is an art to the way we use combat actions in pursuit of our larger objectives. We must view each envisioned action—battle, engagement, interdiction mission, feint, or refusal to give battle—as a element of a larger whole rather than as an independent, self-contained event.

While combat is an integral part of war, it is by nature costly. The flames of war are fueled by money, material stocks, and human lives. As Eisenhower wrote, the word war "is syn-onymous with waste . . . . The problem is to determine how, in time and space, to expend assets so as to achieve the maximum in results."[4] Economy dictates that we use combat actions wisely.

We do this first by fighting when it is to our advantage to do so—when we are strong compared to the enemy or we have identified some exploitable vulnerability—and by avoiding bat-tle when we are at a disadvantage. When we are at a disadvan-tage tactically, economy leads to refusing to engage in battle in that particular situation. When we are at a tactical disadvan-tage theater-wide, it leads to waging a campaign based on hit-and-run tactics and a general refusal to give pitched battle, ex-cept when local advantage exists. This can be seen in countless historical examples: Rome under Fabius versus Hannibal, the Viet Cong in Vietnam, Washington and Nathanael Greene in the Revolutionary War.

By the same token, given a theater-wide tactical advantage, we might want to bring the enemy to battle at every opportu-nity: Rome under Varro versus Hannibal, the United States in Vietnam, Eisenhower in Europe, or Grant versus Lee. Never-theless, such an approach is generally time-consuming, and success depends on three conditions: first, and most important, there is something to be gained strategically by exploiting this tactical advantage as in Grant's series of battles with Lee; sec-ond, popular support for this approach will outlast the enemy's ability to absorb losses as was not the case with the United

States in Vietnam; and third, the enemy is willing or can be compelled to accept battle on a large scale as the Germans were in Europe in 1944, but the Viet Cong generally were not.

It is not sufficient to give battle simply because it is tactically advantageous to do so. It is more important that battle be strategically advantageous or strategically necessary. That is, there should be something to gain by fighting or to lose by not fighting. Strategic gain or necessity can be sufficient reason even when the situation is tactically disadvantageous. Consequently, it is conceivable that we might accept battle even expecting a tactical defeat if the results will serve the goals of strategy. For example, after running away from Cornwallis' British forces in the Carolinas for 6 weeks in 1781, Nathanael Greene could decide to give battle "on the theory that he could hardly lose. If Cornwallis should win a tactical victory, he was already so far gone in exhaustion it would probably hurt him almost as much as a defeat."[5]

Ideally, operational commanders fight only when and where they want to. Their ability to do this is largely a function of their ability to maintain the initiative and shape the events of war to their purposes. "In war it is all-important to gain and retain the initiative, to make the enemy conform to your action, to dance to your tune."[6] Retaining the initiative, in turn, is largely the product of maintaining a higher operational tempo, which we will discuss later in this chapter.

Even so, we must realize that we may not always be able to fight on our own terms. We may be compelled to fight because of strategic constraints (like Lee's requirement to defend Richmond) or by a skillful enemy who perceives an advantage and seeks battle. In such cases, we have no choice but to give battle in a way that serves our strategy to the extent possible and to exploit all possible advantage of the tactical results.

The conduct of a battle, once joined, is principally a tactical problem, but even the tactician should keep larger aims in mind as he fights. As an example, consider General Guderian at the Battle of Sedan in May 1940. (See figure on page 68.) Guderian's XIXth Panzer corps was attacking generally south with the strategic aim "to win a bridgehead over the Meuse at Sedan and thus to help the infantry divisions that would be following to cross that river. No instructions were given as to what was to be done in the event of a surprise success."[7] By 13 May, Guderian had forced a small bridgehead. By the 14th, he had expanded the bridgehead to the south and west but had not broken through the French defenses. Lacking instructions on how to continue the battle, Guderian opted to attack west in concert with the strategic aim of the campaign. "1st and 2nd Panzer Divisions received orders immediately to change direction with all their forces, to cross the Ardennes Canal, and to head west with the objective of breaking clear through the French defenses."[8] Guderian's forces broke through and sped all the way to the coast at the English Channel, cutting off the Anglo-French armies to the north.

Tactics Supporting Operations: Guderian, 1940

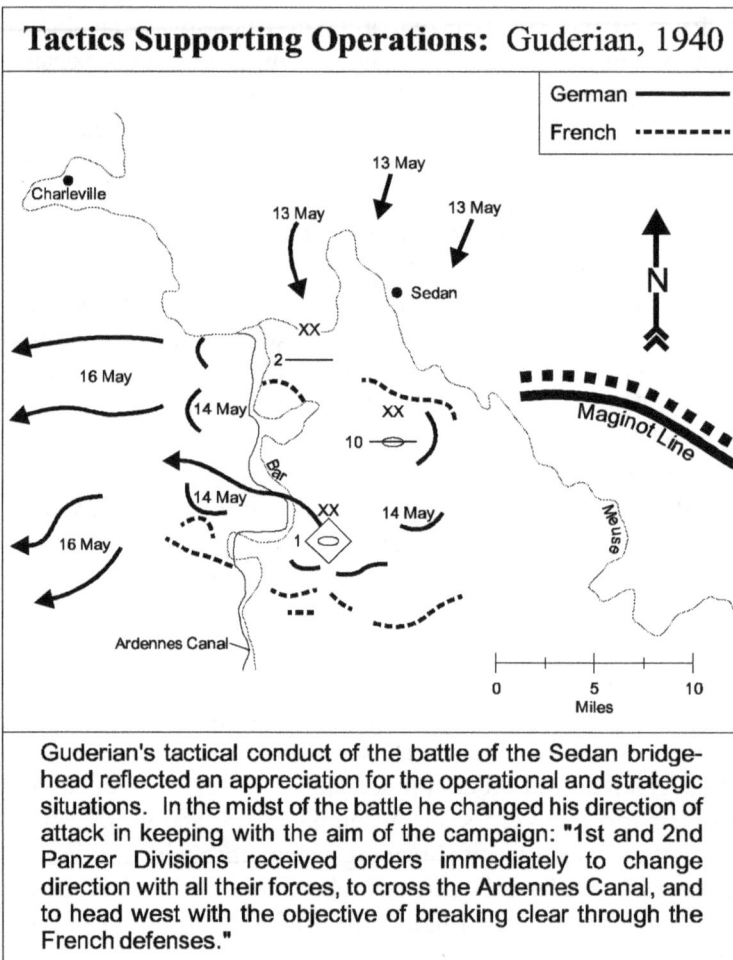

Guderian's tactical conduct of the battle of the Sedan bridge-head reflected an appreciation for the operational and strategic situations. In the midst of the battle he changed his direction of attack in keeping with the aim of the campaign: "1st and 2nd Panzer Divisions received orders immediately to change direction with all their forces, to cross the Ardennes Canal, and to head west with the objective of breaking clear through the French defenses."

PERSPECTIVE

The campaign demands a markedly different perspective than the battle. It requires us to "think big," as Field-Marshal Slim put it, seeing beyond the parameters of immediate combat to the requirements of theater strategy as the basis for deciding when, where, and who to fight. We should view no tactical action in isolation, but always in light of the design for the theater as a whole.

While the tactician looks at the immediate tactical problem and the conditions directly preceding and following, the operational commander must take a broader view. The operational commander must not become so involved in tactical activities as to lose the proper perspective. This broader perspective implies broader dimensions of time and space over which to apply the military art. The actual dimensions of the operational canvas vary with the nature of the war, the size and capabilities of available forces, and the geographical characteristics of the theater. Nonetheless, all the time and space subject to the commander's influence must be considered to create the conditions of success. In 1809, Napoleon carried with him maps of the entire continent of Europe, thereby enabling consideration of operations wherever they suited his purposes. Similarly, Rommel's intervention in the North African theater of war in

1942 successfully delayed American and British efforts to open up a second front in support of their Russian allies.

Based on this larger perspective, the operational commander's concern with military geography is on a different scale than that of the tactical commander. The operational commander is not concerned with the details of terrain that are of critical importance to the tactician in combat, such as hills, draws, fingers, clearings or small woods, creeks, or broken trails. Rather, the operational commander's concern is with major geographical features which can bear on the campaign: rivers and major watersheds, road systems, railways, mountain ranges, urban areas, airfields, ports, and natural resource areas. Patton believed that "in the higher echelons, a layered map of the whole theater to a reasonable scale, showing roads, railways, streams, and towns is more useful than a large-scale map cluttered up with ground forms and a multiplicity of nonessential information."[9] His concern was with the movement of large forces.

We describe activities at the strategic level as bearing directly on the war overall, at the operational level as bearing on the campaign, and at the tactical level as bearing on combat—that is, on the engagement or battle. Therefore, in designing and executing a campaign, we seek to focus on the attainment of strategic and operational objectives. At the same time, we adapt to the realities of the tactical situation.

SURPRISE

Surprise is a state of disorientation that results from unexpected events and degrades the ability to react effectively. Surprise can be of decisive importance. *Tactical surprise* catches the enemy unprepared in such a way as to affect the outcome of combat. It is of a relatively immediate and local nature. *Operational surprise* catches the enemy unprepared in such a way as to impact on the campaign. To achieve operational surprise, we need not necessarily catch the enemy tactically unaware. For example, at the Inchon landing in 1950, the need first to capture Wolmi-do Island, which dominated the inner approaches to Inchon harbor, removed any hope of achieving tactical surprise with the main landings. Operational surprise was nonetheless complete. Even though the assault on Wolmi-do Island was preceded by a 5-day aerial bombardment, the North Korean army, far to the south menacing Pusan, could not react in time. Wolmi-do was cut off and soon collapsed.

Surprise may be the product of *deception* that misleads the enemy into acting in a way prejudicial to his interests.[10] For example, the Normandy invasion succeeded in large part because an elaborate deception plan convinced the Germans that the invasion would take place at Calais. Long after Allied forces were established ashore in Normandy, vital German reserves were held back awaiting the real invasion elsewhere. A major factor in the success of the deception plan was that it was designed to exploit a known enemy belief that General George Patton—in the Germans' opinion the best Allied operational commander—would lead the key attack.[11]

Surprise may also be the product of *ambiguity* when we generate many options and leave the enemy confounded as to which we will pursue. For example, prior to the Allied invasion of North Africa in 1942, Eisenhower's choice of a thousand miles of coastline from Casablanca to Tunis precluded the Axis forces from anticipating the actual landing sites.

Surprise may simply be the product of *stealth* where the enemy is not deceived or confused as to our intentions but is ignorant of them. Exploiting his knowledge of Japanese intentions and their total ignorance of his, Admiral Nimitz was able to strike a decisive blow against the Japanese invasion fleet at the Battle of Midway in June 1942.

Of these three sources of surprise, deception may offer the greatest potential payoff because it deludes the enemy into actions we actively desire him to take. However, because deception means actually convincing the enemy of a lie rather than simply leaving him confused or ignorant, it is also the most difficult to execute. This is even truer at the operational level than at the tactical. Due to the broader perspective of operations, operational deception must feed false informa- tion to a wider array of enemy intelligence collection means over a longer period of time than is the case with tactical deception. This increases the complexity of the deception effort, the need for consistency, and the risk of compromise.

TEMPO

Tempo is a rhythm of activity. It is a significant weapon because it is through a faster tempo that we seize the initiative and dictate the terms of war. *Tactical tempo* is the pace of events within an engagement. *Operational tempo* is the pace of events between engagements. In other words, in seeking to control tempo, we need the ability to shift from one tactical action to another consistently faster than the enemy. Thus it is not in absolute terms that tempo matters, but in terms relative to the enemy.

We create operational tempo in several ways. First, we gain tempo by undertaking multiple tactical actions simultaneously such as the German blitzes into Poland and France in 1939 and 1940 which were characterized by multiple, broadly dispersed thrusts. Second, we gain tempo by anticipating the various likely results of tactical actions and preparing sequels for exploiting those results without delay. Third, we generate tempo by decentralizing decisionmaking within the framework of a unifying intent. Slim recalled of his experience in Burma in the Second World War—

Commanders at all levels had to act more on their own; they were given greater latitude to work out their own plans to achieve what they knew was the Army Commander's intention. In time they developed to a marked degree a flexibility of mind and a firmness of decision that enabled them to act swiftly to take advantage of sudden information or changing circumstances without reference to their superiors.[12]

Finally, we maintain tempo by avoiding unnecessary combat. Any battle or engagement, even if it allows us to destroy the enemy, takes time and energy, and this saps our operational tempo. Here we see another reason besides the desire for economy to fight only when and where necessary. Conversely, by maintaining superior operational tempo, we can lessen the need to resort to combat. The German blitzkrieg through France in 1940 was characterized more by the calculated avoidance of pitched battle after the breakthrough than by great tactical victories. By contrast, French doctrine at the time called for deliberate, methodical battle. When the German tempo of operations rendered this approach impossible to implement, the defenders were overwhelmed. The French were unable to reconstitute an organized resistance and force the Germans to fight for their gains.[13] Liddell Hart wrote of the 1940 campaign in France—

The issue turned on the time factor at stage after stage. French countermeasures were repeatedly thrown out of gear because their timing was too slow to catch up with the changing situations . . . .

> The French commanders, trained in the slow-motion methods
> of 1918, were mentally unfitted to cope with the panzer pace,
> and it produced a spreading paralysis among them.[14]

As with almost everything at the operational level of war, controlling the tempo of operations requires not only speed, but a solid understanding of the operational and strategic goals of the campaign. During Desert Storm, for instance, the Marine Corps' drive on the main effort's right flank rolled forward much faster than higher commanders had anticipated. Although this fast pace unquestionably offered tactical advantages within the Marines' area of operations, from the standpoint of the overall Allied plan it posed problems. Rather than fixing the Iraqi forces in place, as planned, the Marines were routing them. This created the possibility that major Iraqi forces would flee the trap before other Allied forces could close the envelopment from the left. Had the primary objective been the destruction of the Iraqi army, it might have been necessary to slow the Marines' advance even though this might have increased their casualties in the long run. The main objective, however, was to free Kuwait of Iraqi occupation. Given that the Iraqis had already broken and started running, there was no guarantee that slowing the tempo on the right would have the desired effect. Therefore, the wisest course—and the one that was taken—was to let the Marines maintain their high tempo, while expediting the movements of other Allied formations.[15]

SYNERGY

The conduct of a successful campaign requires the integration of many disparate efforts. Effective action in any single war-fighting function is rarely decisive in and of itself. We obtain maximum impact when we harmonize all warfighting functions to accomplish the desired strategic objective in the shortest time possible and with minimal casualties.[16] Within the context of the campaign, we focus on six major functions: command and control, maneuver, fires, intelligence, logistics, and force protection.[17]

Command and Control

No single activity in war is more important than command and control. Without command and control, military units degenerate into mobs, the subordination of military force to policy is replaced by random violence, and it is impossible to conduct a campaign. Command and control encompasses all military operations and functions, harmonizing them into a meaningful whole. It provides the intellectual framework and physical structures through which commanders transmit their intent and decisions to the force and receive feedback on the results. In short, command and control is the means by which a commander recognizes what needs to be done and sees to it that appropriate actions are taken.[18]

Command and control during the conduct of a campaign places unique requirements on the commander, the command

and control organization, and the command and control support structure. The scope of activities in the campaign (both in time and space) will likely be vastly greater than in a battle or engagement. The number of organizational players will also influence the effective conduct of command and control. In any modern campaign, the commander must be concerned with more than just the higher headquarters and subordinate elements. A wide range of participants must be informed and coordinated with, both military (such as other units of a joint or multinational force) and civilian (such as other governmental agencies, host nation authorities, and nongovernmental organizations). Information management is a key function since communications and information systems can generate a flood of information. It is important to ensure that this flood of information does not overwhelm us but provides meaningful knowledge to help reduce uncertainty. Finally, the nature of these factors can make it difficult to ensure that the commander's intent and decisions are understood throughout the force and implemented as desired.

In implementing command and control during the campaign, we seek to reduce uncertainty, facilitate decisionmaking, and help generate a high operational tempo. Through effective information management and a well-designed command and control support structure, we attempt to build and share situational awareness. Planning is another essential element of command and control. Campaign design is largely the result of planning, and planning continues throughout the campaign as the campaign plan is modified and adapted based upon the changes in the situation and the results of campaign activities. We must prepare to function or even thrive in an environment of uncertainty and to make decisions despite incomplete or unclear information. A clear statement of intent that is understood throughout the force, flexible plans, an ability to adapt to unforeseen circumstances, and the initiative to recognize and seize opportunities as they present themselves permit us to generate tempo and perform effectively despite uncertainty.

Maneuver

Maneuver is the movement of forces for the purpose of gaining an advantage over the enemy in order to accomplish our objectives. While tactical maneuver aims to gain an advantage in combat, operational maneuver seeks to gain an advantage bearing directly on the outcome of the campaign or in the theater as a whole.

A classic example of operational maneuver was General MacArthur's landing at Inchon in 1950. (See figure.) The bulk of North Korea's army was well to the south, hemming the U.S. Eighth Army into the Pusan perimeter. Using the sea as maneuver space, MacArthur conducted a classic turning movement. By landing X Corps at Inchon, MacArthur threatened the enemy's lines of communications and forced the overextended enemy to shift fronts. This maneuver not only cut the North Koreans' flow of supplies and reinforcements but also forced them to move in a way that exposed them to a counterattack from the south.

Operational Manuever: MacArthur, 1950

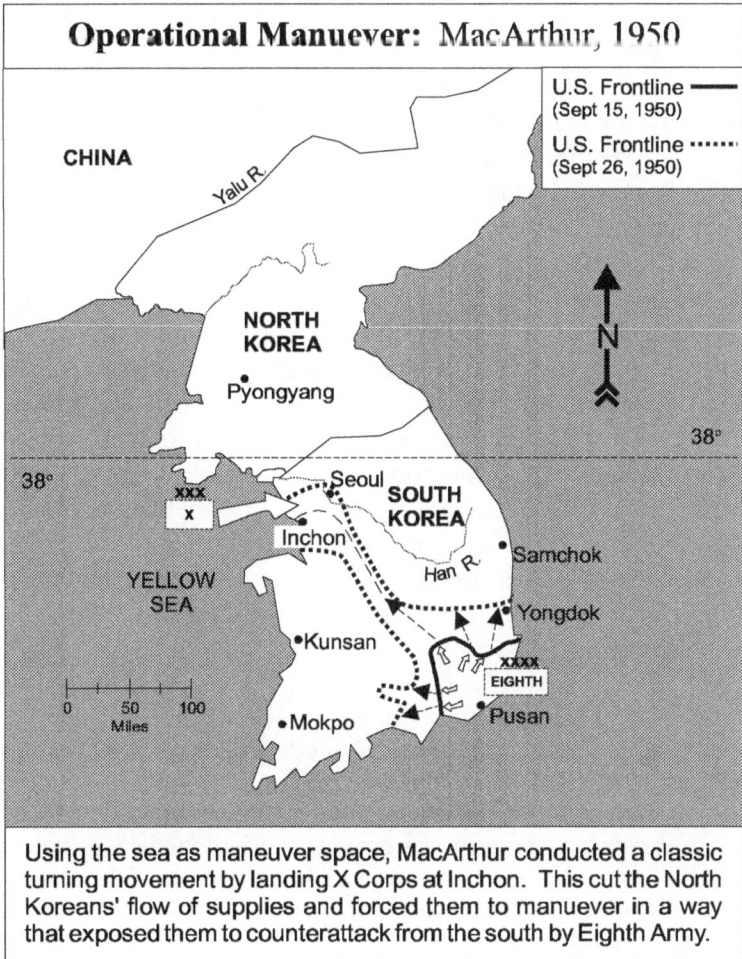

Using the sea as maneuver space, MacArthur conducted a classic turning movement by landing X Corps at Inchon. This cut the North Koreans' flow of supplies and forced them to manuever in a way that exposed them to counterattack from the south by Eighth Army.

Operational maneuver allows us to create and to exploit opportunities. It affords us the opportunity to develop plans which employ multiple options, or branches.[19] A branch plan helps us to anticipate future actions. Operational maneuver provides the means by which we can assess the situation, determine the branch which offers the best opportunity for success, and implement the decision. By skillful use of branches, we add to our flexibility and speed.

General Sherman's campaign in Georgia in 1864 illustrates the use of operational maneuver to retain the initiative and keep the opposition off balance. (See figure.) During his march through Georgia, Sherman ingeniously sought to keep his opponent constantly on the horns of a dilemma. His line of advance kept the Confederates in doubt whether his next objective was first Macon or Augusta, and then Augusta or Savannah. Sherman was ready to take whichever objective conditions favored. Campaigning through the Carolinas Sherman repeated this approach—

> so that his opponents could not decide whether to cover Augusta or Charleston, and their forces became divided. Then after he had ignored both points and swept between them to gain Columbia . . . the Confederates were kept in uncertainty as to whether Sherman was aiming for Charlotte or Fayetteville. [Finally, when] he advanced from Fayetteville they could not tell whether Raleigh or Goldsborough was his next, and final, objective.[20]

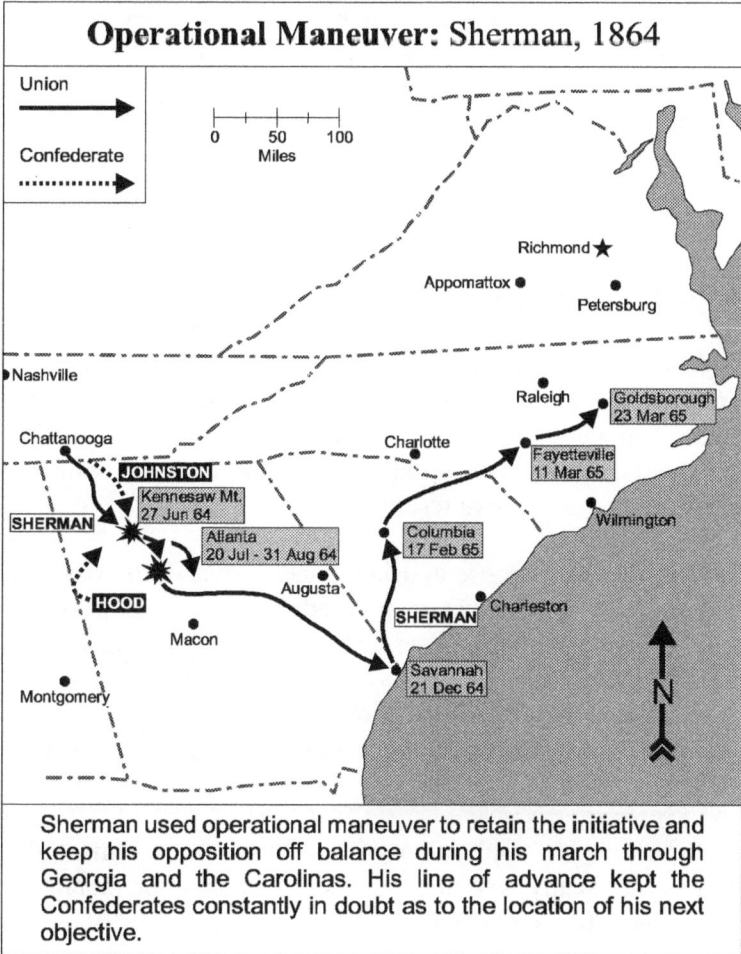

Operational Maneuver: Sherman, 1864

Union

Confederate

0   50   100
        Miles

Richmond ★

Appomattox ●

Petersburg

● Nashville

Chattanooga

Raleigh

Goldsborough
23 Mar 65

JOHNSTON

Charlotte

Fayetteville
11 Mar 65

Kennesaw Mt.
27 Jun 64

SHERMAN

Wilmington

Atlanta
20 Jul - 31 Aug 64

Columbia
17 Feb 65

HOOD

Augusta

SHERMAN

Charleston

Macon

Savannah
21 Dec 64

Montgomery

N

Sherman used operational maneuver to retain the initiative and keep his opposition off balance during his march through Georgia and the Carolinas. His line of advance kept the Confederates constantly in doubt as to the location of his next objective.

If tactical maneuver takes place during and within battle, operational maneuver takes place before, after, and beyond battle. The operational commander seeks to secure a decisive advantage before the battle is joined by rapid, flexible, and opportunistic maneuver. Such action allows us to gain the initiative and shape the action to create a decisive advantage.

The operational commander also uses maneuver to exploit tactical success, always seeking to achieve strategic results. The commander must be prepared to react to the unexpected and exploit opportunities created by conditions which develop from the initial action. By exploiting opportunities, we create in increasing numbers more opportunities for exploitation. The ability and willingness to ruthlessly exploit these opportunities often generates decisive results.

Our ultimate purpose in using maneuver is not to avoid battle, but to give ourselves such an advantage that the result of the battle is a matter of course. In the words of Liddell Hart, the *"true aim is not so much to seek battle as to seek a strategic situation so advantageous that if it does not of itself produce the decision, its continuation by a battle is sure to achieve this."*[21]

If the classic application of maneuver is movement that places the enemy at a disadvantage, then superior mobility—the capability to move from place to place faster than the enemy while retaining the ability to perform the mission—is a key ingredient of maneuver. The object is to use mobility to gain an advantage by creating superiority at the point of battle or to avoid disadvantageous battle altogether. [22]

Operational mobility is the ability to move between engagements and battles within the context of the campaign. It is a function of range and sustained speed over distance.[23] Patton recognized the importance of distinguishing between tactical and operational mobility when he wrote: "Use roads to march on; fields to fight on . . . when the roads are available for use, you save time and effort by staying on them until shot off."[24] If the essence of the operational level is deciding when and where to fight, operational mobility is the means by which we commit the necessary forces based on that decision.

An advantage in operational mobility can have a significant impact. In the Second World War in the Pacific island-hopping campaign, the Allies used operational mobility that allowed them to shift forces faster than the Japanese. The result was that Japanese forces were cut off and allowed to wither while the Allies consistently moved towards the Japanese home islands to bring them under direct attack.

Although we typically think of shipping as an element of strategic mobility, it may be employed to operational effect as well. In many cases, an amphibious force can enjoy greater operational mobility moving along a coastline than an enemy moving along the coast by roads, particularly when the amphibious force has the ability to interfere with the enemy's use of those roads. The same use can be made of airlift. Such an advantage in operational mobility can be decisive.

Fires

We employ fires to delay, disrupt, degrade, or destroy enemy capabilities, forces, or facilities as well as to affect the enemy's will to fight. Our use of fires is not the wholesale attack of every unit, position, piece of equipment, or installa- tion we find. Rather, it is the selective application of fires to reduce or eliminate a key element, resulting in a major disabling of the enemy system. We use fires in harmony with maneuver against those enemy capabilities, the loss of which can have a decisive impact on the campaign or major operation.

During the conduct of the campaign, we use fires to shape the battlespace. By shaping, we influence events in a manner which changes the general condition of war decisively to our advantage. "Shaping activities may render the enemy vulnerable to attack, facilitate maneuver of friendly forces, and dictate the time and place for decisive battle."[25] Through those actions, we gain the initiative, preserve momentum, and control the tempo of the campaign. Operation Desert Storm provides an excellent example of a successful shaping effort. Our extensive air operations destroyed facilities, eliminated the Iraqi navy and air force, reduced the effectiveness of ground forces within Kuwait, and shattered the enemy's cohesion. An elaborate deception plan also confused the Iraqis as to the size and location of ground attacks while intense psychological operations helped undermine their morale. The end result was an enemy who was both physically and mentally incapable of countering the maneuver of Coalition forces.

Campaign planners must analyze the enemy's situation, keeping in mind the commander's mission, objectives, intent, and our capabilities available for employment. We seek to target those enemy vulnerabilities that, if exploited, will deny resources critical to the enemy's ability to resist.[26] These targets may range from military formations, weapon systems, or command and control nodes to the target audiences for a psychological operation. However, the nature of these targets is situationally dependent and is based on an analysis of the enemy and our mission.

Intelligence

Intelligence is crucial to both the design and conduct of the campaign. Intelligence underpins the campaign design by providing an understanding of the enemy and the area of operations as well as by identifying the enemy's centers of gravity and critical vulnerabilities. During the conduct of the campaign, intelligence assists us in developing and refining our understanding of the situation, alerts us to new opportunities, and helps to assess the effects of actions upon the enemy. Intelligence cannot provide certainty; uncertainty is an inherent attribute of war. Rather, intelligence estimates the possibilities and probabilities in an effort to reduce uncertainty to a reasonable level.

Because the operational level of war aims to attain a strategic objective through the conduct of tactical actions, operational intelligence must provide insight into both the strategic and tactical situations as well as all factors that influence them. The differences among the tactical, operational, and strategic levels of intelligence lie in the scope, application, and level of detail associated with each level. *Operational intelligence* pertains broadly to the location, capabilities, and intentions of enemy forces that can conduct campaigns or major operations. It also is concerned with all operational aspects of the environment that can impact on the campaign such as geography, the national or regional economic and political situation, and fundamental cultural factors. Operational intelligence is less concerned with individual enemy units than it is with major formations and groupings. Similarly, it concentrates on general aspects of military geography such as mountain ranges or river valleys rather than on individual pieces of key terrain or a specific river-crossing site. Operational intelligence should be focused on patterns of activity, trends, and indications of future intentions. It should examine the enemy as a system rather than as individual components in an effort to determine how the entire enemy organization functions and as a means to identify the enemy's strengths, weakness, centers of gravity, and critical vulnerabilities.

During the execution of the campaign plan, intelligence strives to provide as detailed and accurate a picture of the current situation as possible while updating the estimate of the enemy's capabilities and intentions. Intelligence is a key ingredient in gaining and maintaining situational awareness and makes an essential contribution to the conduct of the campaign through its support to targeting, force protection, and combat assessment. Intelligence operations are conducted throughout the campaign. Just as campaign plans are based on intelligence, intelligence plans are grounded in operations. The intelligence collection, production, and dissemination efforts are integrated with planned operations to support modification of ongoing activities, execution of branches and sequels, exploitation of success, and shaping the battlespace for future operations.

The successful use of intelligence at the operational level was illustrated in the dramatic victory achieved by U.S. naval forces in the Battle of Midway in June 1942. Japanese naval successes during the months following their attack on Pearl Harbor had provided them enormous advantage. In particular, their significant aircraft carrier strengths provided them with tactical warfighting capabilities far superior to those of the Allies. The questions facing Admiral Nimitz, Commander-in-Chief, U.S. Pacific Fleet, were: What would the Japanese do next? Would they continue, and if so, where?

Intelligence helped provide the answer. U.S. naval intelligence succeeded in breaking the codes used by the Japanese fleet to encrypt radio messages. The resulting intelligence reports, codenamed "Magic," provided significant insight into Japanese operations. Analysis of Magic reports combined with other intelligence uncovered the Japanese intentions to strike at Midway in early June. Using this intelligence to obtain an

operational advantage, Nimitz concentrated his numerically inferior forces where they could ambush the main body of the
Japanese invasion fleet. U.S. forces achieved complete surprise
and sank four Japanese carriers. Their overwhelming success
in defeating a numerically superior enemy proved to be the major turning point in the Pacific theater of operations, dramati-
cally altering the balance of naval power in a single decisive
engagement.[27]

Logistics

At the operational level much more than at the tactical, logistics dictates what is possible and what is not. "A campaign
plan that cannot be logistically supported is not a plan at all,
but simply an expression of fanciful wishes."[28]

Logistics encompasses all activities required to move and
sustain military forces.[29] *Strategic logistics* involves the acquisition and stocking of war materials and the generation and
movement of forces and materials to various theaters. At the
opposite end of the spectrum, *tactical logistics* is concerned
with sustaining forces in combat. It deals with the feeding and
care, arming, fueling, maintaining, and movement of troops
and equipment. In order to perform these functions, the tactical
commander must be provided the necessary resources.

Operational logistics links the strategic source of the means
of war to its tactical employment.[30] During campaign execution, the focus of the logistics effort is on the provision of resources necessary to support tactical actions and the

management of resources to sustain operations throughout the course of the campaign.

The provision of resources to the tactical forces requires a procurement of necessary material as well as the creation and maintenance of an effective theater transportation system. Procurement is usually accomplished through the strategic logistics system. However, when capabilities or assets cannot be obtained from strategic-level sources, our logistics system must be able to obtain the necessary support from host nation, allied, or other sources. The transportation system must have sufficient capacity and redundancy to sustain the necessary level of effort. Transportation requires sufficient ports of entry to receive the needed volume of resources, adequate means of storage, and lines of communications (land, sea, and air) sufficient to move those resources within the theater of operations.

Managing the often limited resources necessary to implement the commander's concept and to sustain the campaign is just as important as providing and delivering the resources to the tactical commanders. At the operational level, logistics demands an appreciation for the expenditure of resources and the timely anticipation of requirements. This requires both the apportioning of resources among tactical forces based on the operational plan and the rationing of resources to ensure sustainment throughout the duration of the campaign. While failure to anticipate logistical requirements at the tactical level can result in delays of hours or days, the same failure at the operational level can result in delays of weeks. Such delays can be extremely costly.

Finally, the provision of logistics in conduct of the campaign demands adaptability. We expect our plans to change. Flexibility in planning and organization coupled with the logistician's continuous situational awareness can foster the innovation and responsiveness necessary to meet these challenges. A dramatic example of adaptability in the provision of logistics occurred during Operation Desert Storm. Just before the start of offensive ground operations, a change in the Marine Forces' concept of operations created the requirement to reposition a significant portion of the logistics support structure. Early recognition of the requirement and flexibility of organization permitted the reconfiguration of support capabilities and the timely movement of necessary resources. An immense hardened forward staging base covering over 11,000 acres was constructed in just 14 days. Fifteen days of ammunition for two divisions; 5 million gallons of petroleum, oils, and lubricants; a million gallons of water; and the third largest naval hospital in the world were positioned before the assault.[31]

Force Protection

We need to take every possible measure to conserve our forces' fighting potential so that it can be applied at the decisive time and place. We accomplish this through properly planning and executing force protection. These actions imply more than base defense or self-protection procedures. At the operational level, force protection means that we must plan to frustrate the enemy's attempts to locate and strike our troops, equipment, capabilities, and facilities. Force protection actions may also extend to keeping air, land, and sea lines of communications free from enemy interference.

Force protection safeguards our own centers of gravity and protects, conceals, reduces, or eliminates critical vulnerabilities. When we are involved in military operations other than war, force protection may include the additional task of protecting the supported nation's population, infrastructure, and economic or governmental institutions. Force protection also encompasses taking precautions against terrorist activities against our own forces and noncombatants.

Successful force protection begins with the determination of indicators that might reveal our plans and movements to enemy intelligence systems. By identifying these indicators and then taking appropriate steps to reduce or eliminate them, we can significantly decrease the potential for the enemy to disrupt our operations.

Aggressive force protection planning and execution improves our ability to maneuver against the enemy and to achieve our operational objectives. By safeguarding centers of gravity, protecting our troops and equipment, and ensuring the security of our installations and facilities, we conserve our combat power so that it can be applied at a decisive time and place.

LEADERSHIP

Leadership is the ability to get human beings to put forth their efforts in pursuit of a collective goal. Strong leadership creates an understanding of goals and a strong commitment to them among all members of the organization. At the higher levels of command, leadership is much less a matter of direct personal example and intervention than it is a matter of being able to energize and unify the efforts of large groups of people, sometimes dispersed over great distances.

This is not to say that personal contact is unimportant at the operational level, nor that charisma and strength of personality do not matter. In fact, we might argue that an operational commander who must influence more people spread over greater distances must be correspondingly more charismatic and stronger of personality than the tactical command- er. The commander must see and be seen by subordinates. As the Supreme Commander in Europe, Eisenhower spent a great deal of time traveling throughout the theater partly to see and to be seen by his men. Nor does this imply that the operational commander does not intervene in the actions of subordinates when necessary. Just as planning at the operational level requires leaders who can decide when and where to fight, campaign execution requires leaders who can determine when and where to use personal influence.

Leadership at the operational level requires clarity of vision, strength of will, and great moral courage. Moreover, it requires the ability to communicate these traits clearly and powerfully through numerous layers of command, each of which adds to the friction inhibiting effective communication. British Field-Marshal Sir William Slim, who in early 1945 retook Burma from the Japanese in a brilliant jungle campaign, noted this requirement by saying that the operational commander must possess "the power to make his intentions clear right through the force."[32]

Operational commanders must establish a climate of cohesion among the widely dispersed elements of their commands and with adjacent and higher headquarters as well. Because they cannot become overly involved in tactics, operational commanders must have confidence in their subordinate commanders. With these subordinates, commanders must develop a deep mutual trust. They must also cultivate in subordinates an implicit understanding of their own operating style and an explicit knowledge of their specific campaign intent. Operational commanders must train their staffs until the staffs become extensions of the commanders' personality.

The nature of campaigns places heavy demands on a leader's communications skills, demands that are quite different from those experienced by tactical unit commanders. Operational commanders must coordinate units from other services and nations. Operational commanders must maintain effective relationships with external organizations, which is particularly difficult when other cultures are involved. Operational commanders must be able to win consensus for joint or multinational concepts of operations and represent effectively to higher headquarters the capabilities, limitations, and external support requirements of their forces.

Conclusion

"Those who know when to fight and when not to fight are victorious. Those who discern when to use many or few troops are victorious. Those whose upper and lower ranks have the same desire are victorious. Those who face the unprepared with preparation are victorious."[1]

—Sun Tzu

A t the risk of belaboring a point, we will repeat for the last time that tactical success of itself does not necessarily bring strategic success. "It is possible to win all the battles and still lose the war. If the battles do not lead to the achievement of the strategic objective, then, successful or not, they are just so much wasted effort."[2] Strategic success that attains the objectives of policy is the military goal in war. Thus we recognize the need for a discipline of the military art that synthesizes tactical results to create the military conditions that induce strategic success. We have discussed the campaign as the principal vehicle by which we accomplish this synthesis.

Understandably perhaps, as tactics has long been a Marine Corps strength, we tend to focus on the tactical aspects of war to the neglect of the operational aspects. This neglect may be also caused by the often contradictory virtues of the two levels: the headlong tactical focus on winning in combat (and the spoiling-for-a-fight mentality it necessarily promotes) compared to the operational desire to use combat sparingly. As we have seen, actions at the higher levels in the hierarchy of war tend to overpower actions at the lower levels, and neglect of the operational level can prove disastrous even in the face of tactical competence. Without an operational design which synthesizes tactical results into a coalescent whole, what passes for operations is simply the accumulation of tactical victories.

Tactical competence can rarely attain victory in the face of operational incompetence, while operational ignorance can squander what tactical hard work has gained. As the price of war is human lives, it is therefore incumbent upon every commander to attain the objective as economically as possible. Operational leaders must understand strategic issues and the fundamentally political nature of all strategic goals. The design and conduct of a successful campaign results from a clear understanding of the relationship between strategic and operational objectives, the interaction between the military and other instruments of national power, and the need for judicious and effective use of combat to achieve the objectives.

The Campaign

1.   Henri Jomini, *The Art of War* (Westport, CT: Greenwood Press, 1971) p. 178. What Jomini describes as strategic would be classified as operational by today's construct.

2.   B. H. Liddell Hart, *Strategy* (New York: Praeger, 1967) p. 338.

3.   *The Memoirs of Field-Marshal Montgomery* (New York: World Publishing Co., 1958) p. 197.

4.   Joint Pub 1-02, *Department of Defense Dictionary of Military and Associated Terms.*

5.   Joint Pub 3-0, *Doctrine for Joint Operations* (February 1995) p. III-4.

6.   Liddell Hart, *Strategy*, p. 351.

7.   **Military strategy:** "The art and science of employing the armed forces of a nation to secure the objectives of national policy by the application of force or the threat of force."
     Strategic level of war: "The level of war at which a nation, often as a member of a group of nations, determines national or multinational (alliance or coalition) security objectives and guidance, and develops and uses national resources to accomplish these objectives. Activities at this level establish national and multinational military objectives; sequence initiatives; define limits and assess risks for the use of military and other instruments of national

power; develop global plans or theater war plans to achieve these objectives; and provide military forces and other capabilities in accordance with strategic plans." (Joint Pub 1-02)

8. **Strategic concept:** "The course of action accepted as the result of the estimate of the strategic situation. It is a statement of what is to be done in broad terms sufficiently flexible to permit its use in framing the military, diplomatic, economic, psychological and other measures which stem from it." (Joint Pub 1-02) Sometimes itself referred to as a "strategy."

9. **Tactical level of war:** "The level of war at which battles and engagements are planned and executed to accomplish military objectives assigned to tactical units or task forces. Activities at this level focus on the ordered arrangement and maneuver of combat elements in relation to each other and to the enemy to achieve combat objectives." (Joint Pub 1-02)

10. **Operational level of war:** "The level of war at which campaigns and major operations are planned, conducted, and sustained to accomplish strategic objectives within theaters or areas of operations. Activities at this level link tactics and strategy by establishing operational objectives needed to accomplish the strategic objectives, sequencing events to achieve the operational objectives, initiating actions, and applying resources to bring about and sustain these events. These activities imply a broader dimension of time or space than do tactics; they ensure the logistic and administrative support of tactical forces, and provide the means by which tactical successes are exploited to achieve strategic objectives." (Joint Pub 1-02)

11. Erich von Manstein, *Lost Victories* (Novato, CA: Presidio Press, 1982) p. 79.

12. David Jablonsky, "Strategy and the Operational Level of War," *The Operational Art of Warfare Across the Spectrum of Conflict* (Carlisle, PA: U.S. Army War College, 1987) p. 11.

13. In fact, they can be quite small. For example, consider the killing of Haitian guerrilla leader Charlemagne Peralte by two Marine noncommissioned officers in 1919. During this period, U.S. Marines were involved in the occupation of Haiti. Peralte had raised a rebel force of as many as 5,000 in the northern part of the country. From February through October, Marine forces pursued the rebels, known as "cacos," fighting 131 engagements but were unable to suppress the rebel activity. So, disguised as cacos, Sgt. Herman Hanneken and Cpl. William Button infiltrated Peralte's camp, where Hanneken shot and killed the caco leader. The rebellion in the north subsided. In this case, a special operation consisting of two Marines accomplished what 7 months of combat could not.

14. Carl von Clausewitz, *On War*, trans. and ed., Michael Howard and Peter Paret (Princeton, NJ: Princeton University Press, 1984) p. 607. "No other possibility exists, then, than to subordinate the military point of view to the political."

15. Sun Tzu, *The Art of War*, trans. Samuel B. Griffith (New York: Oxford University Press, 1971) p. 93.

16. Edward N. Luttwak, *Strategy: The Logic of War and Peace* (Cambridge, MA: Harvard University Press, 1987) pp. 69–71 and 208–230 discusses this "interpenetration" of the levels of war.

17. Joint Pub 3-0, *Doctrine for Joint Operations* (February 1995) pp. III-4–III-5.

18. Col W. Hays Parks, U.S. Marine Corps Reserve, "Crossing the Line," *Proceedings* (November 1986) pp. 40–52 and LCdr Joseph T. Stanik, U.S. Navy (Retired), "Welcome to El Dorado Canyon," *Proceedings* (April 1996) pp. 57–62.

19. **Battle:** "A series of related tactical engagements that last longer than an engagement, involve larger forces, and could affect the course of the campaign. They occur when division, corps, or army commanders fight for significant objectives." MCRP 5-2A, *Operational Terms and Graphics* (June 1997).

20. **Engagement:** "A small tactical conflict, usually between opposing maneuver forces." (MCRP 5-2A)

21. For a detailed discussion of the Guilford Courthouse battle and its impact on British operations, see Thomas E. Baker, *Another Such Victory: The Story of the American Defeat at Guilford Courthouse That Helped Win the War of Independence* (New York: Eastern Acorn Press, 1992).

22. Russell F. Weigley, *The American Way of War* (Bloomington, IN: Indiana University Press, 1973) pp. 32–35.

23. Liddell Hart, *Strategy*, p. 338.

24. See J. F. C. Fuller, *Grant and Lee: A Study in Personality and Generalship* (Bloomington, IN: Indiana University Press, 1982) particularly pp. 242–283.

25. Weigley, p. 92.

26. Ibid., p. 118. Fuller, p. 253.

27. Ulysses S. Grant, *Personal Memoirs of U.S. Grant* (New York: Da Capo Press, Inc., 1982) p. 369.

28. Ibid., p. 367. Grant to Sherman, 4 April 1864. Grant repeated this phrase directly from Lincoln.

29. Ibid., p. 366.

30. Fuller, pp. 79–80.

31. Weigley, p. 139.

32. Ibid., p. 108.

33. Ibid., p. 123.

34. Grant, p. 384.

35. Fuller, p. 268. "In this respect there is no difference between Grant and Lee; neither understood the full powers of the rifle or the rifled gun; neither introduced a single tactical innovation of importance, and though the rifle tactics of the South were superior to those of the North, whilst the artillery tactics of the North were superior to those of the South, these differences were due to circumstances outside generalship."

Designing the Campaign

1.   Clausewitz, p. 182.

2.   Liddell Hart, *Strategy*, p. 343–344.

3.   John F. Meehan III, "The Operational Trilogy," *Parameters* (September 1986) p. 15.

4.   The distinction between strategies of annihilation and erosion is discussed further in MCDP 1-1, *Strategy*. It originates in Clausewitz's distinction between limited and unlimited war. See Clausewitz, *On War*, Book I. It was further developed in the theories of Hans Delbrück, whom John Keegan calls "the figure who bestrides the military historian's landscape." John Keegan, *The Face of Battle* (New York: The Viking Press, 1976) p. 53, see also pp. 34–35 and 54. See Hans Delbrück, *History of the Art of War Within the Framework of Political History*, trans. Walter J. Renfroe, Jr. (Westport, CT: Greenwood Press, 1985) especially vol. 3, book III, chapter IV, "Strategy," pp. 293–318.

5.   We use the terms "annihilation" and "incapacitation" more or less interchangeably, but the words themselves pose some problems. Soldiers tend to think of annihilation as the absolute *physical* destruction of all of the enemy's troops and equipment. This is rarely achieved and seldom necessary. Germany still had large numbers of well-armed troops at the end of World War II, yet there is little argument that the *Wehrmacht* was strategically annihilated. Nonmilitary people tend to confuse the military goal of annihilating the enemy's military capacity with political or ideological goals like genocide and extermination and may thus be

shocked or horrified to hear of our plan of annihilation. "Incapacitation," on the other hand, is literally exactly what we mean to convey: the destruction of the enemy's military *capacity* to resist.
Unfortunately, the word also connotes the use of nonlethal weapons
and other limited forms of warmaking that contradict the strategic
concept we seek to convey. To deal with such semantic problems,
military leaders must understand the underlying concepts and, in
describing their strategies, use words appropriate to the particular
audience.

6.   As with annihilation and incapacitation, labels carry some
problems. Attrition has developed a negative connotation because of
the experience of *tactical* attrition in Grant's later campaigns, the
Western Front in World War I, and U.S. actions in Vietnam. Erosion carries no such negative connotations, and that is why we use
it more prominently here. The words mean literally the same thing,
however, and attrition is the traditional term used in classical military theory to describe the concept we wish to convey.

7.   The United States pursued such unlimited political aims in
the American Civil War, World War I, World War II, Grenada in
1983, and Panama in 1989. Another successful example is the
North Vietnamese war against South Vietnam. Unsuccessful examples are the German invasion of Russia in 1940, the North Korean
campaign against South Korea in 1950, and the Russian war
against Chechnya in the mid-1990s.

8.   Some readers will object that this could not have been a
strategy of annihilation because we left Saddam Hussein in power.
That is confusing the political with the military goal. Had we
wished to pursue the overthrow of Saddam's government, we were
well positioned to do so, having eliminated Saddam's air and naval

power, thoroughly demoralized his army, and completely isolated him from external support.

9.   In annihilation strategies, military forces always represent the main effort—with the important exception of internal wars. Such internal struggles for power are very often zero-sum events in which one side's victory entails the other's elimination. Therefore, the opponents seek each other's complete destruction, which normally cannot be achieved until the enemy's military protection is removed. Remember, however, that every government at war has to take political action to maintain the "home front," as well as military action against the enemy. In internal wars, the opponents share a *common* home front. Therefore, economic, diplomatic, and psychological programs (e.g., land reform, political reform, pacification operations, etc.) sometimes take precedence over purely military operations even when the military goal remains annihilation. In Vietnam, for example, the U.S. and the government of South Vietnam waged a strategy of erosion against what they perceived to be an external foe, North Vietnam. Within South Vietnamese borders, however, they waged a war of annihilation against the Viet Cong and the North Vietnamese regulars who supported them. Energetic search-and-destroy and aerial bombing operations against enemy military forces often conflicted with various internal nation-building efforts which sought to create legitimacy for the government in Saigon. The failure to harmonize both military and nonmilitary actions at the operational level often proved counterproductive.

10.   The term "center of gravity," as it is used in military doctrine, originated with Clausewitz. He used the term (*Schwerpunkt* in the original German) in many different ways, usually as a handy metaphor rather than a well-defined doctrinal term. Often he used it

merely to mean "the main thing" or "the most important concern." Our definition derives from a few specific discussions in *On War*, especially pp. 485–486 (which deals with the concept at the operational level) and 595–597 (which looks at the concept in strategic terms). The purpose of identifying centers of gravity (preferably reducing the list to one crucial item) is to force us to think through the essential elements of a particular enemy's power and thus to help us focus on what makes him dangerous and what we need to do to defeat him. Unfortunately, this sometimes leads us into thinking that we must directly attack those strengths. The philosophy of *Warfighting* therefore uses the concept of the critical vulnerability, which forces us to think through creative ways of undermining the enemy's strength at the minimum possible cost and risk to ourselves.

11.   Charles XII of Sweden did in fact lose his army in Russia in 1709 and is considered a failure.

12.   Clausewitz, p. 596.

13.   Ibid., p. 163.

14.   Ibid., p. 77.

15.   Winston S. Churchill, *The World Crisis,* vol. 2 (New York: Charles Scribner's Sons, 1923) p. 5.

16.   Dwight D. Eisenhower, *Crusade in Europe* (New York: Doubleday, 1990) p. 256.

17. L. D. Holder, "Operational Art in the U.S. Army: A New Vigor," *Essays on Strategy*, vol. 3 (Washington, D.C.: National Defense University Press, 1986) p. 124.

18. Eisenhower, p. 176. Also: "In committing troops to battle there are certain minimum objectives to be attained, else the operation is a failure. Beyond this lies the realm of reasonable expectation, while still further beyond lies the realm of hope—all that might happen if fortune persistently smiles upon us.

"A battle plan normally attempts to provide guidance even into this final area, so that no opportunity for extensive exploitation may be lost . . . ." p. 256.

19. These two approaches are also called "progressive" and "inverse." The concept is discussed in the Advanced Amphibious Study Group's, *Planner's Reference Manual* (Draft), vol. 1 (Washington, D.C.: Headquarters, U.S. Marine Corps, 1983) pp. 7-1-6.

20. Chairman Joint Chiefs of Staff Manual 3500.04A, *Universal Joint Task List*, version 3.0 (September 1996).

21. The Confederates understood this too. James M. McPherson, *Battle Cry of Freedom: The Civil War Era* (New York: Ballantine Books, 1989) p. 766.

22. **Reconstitution:** "Those actions that commanders plan and implement to restore units to a desired level of combat effectiveness commensurate with mission requirements and available resources. Reconstitution operations include regeneration and reorganization." (MCRP 5-2A)

23. Eisenhower, p. 228.

24. Ibid., p. 225.

25. Ibid., pp. 228–229.
26. Ibid., p. 229.

27. The Malaysian campaign illustrates the exception noted in footnote 9 on page 105: In internal wars, even a military strategy of annihilation may require the subordination of the military effort to other instruments of power.

28. Joint Pub 5-0, *Doctrine for Planning Joint Operations* (April 1995) pp. II-18–II-21 and MCDP 5, *Planning,* pp. 18–21.

29. Meehan, p. 15.

Conducting the Campaign

1. Clausewitz, p. 77.

2. Quoted in Robert D. Heinl, Jr., *Dictionary of Military and Naval Quotations* (Annapolis, MD: U.S. Naval Academy, 1978) p. 1.

3. Clausewitz, p. 128.

4. Eisenhower, p. 119.

5. Weigley, p. 32. Such a victory is called a "Pyrrhic victory," after the Greek king Pyrrhus. After meeting the Romans in battle for the first time and winning but suffering great losses in the

process, Pyrrhus reportedly said, " 'One more such victory and I am lost.' " R. Ernest Dupuy and Trevor N. Dupuy, *The Encyclopedia of Military History From 3500 B.C. to the Present* (New York: Harper & Row, 1977) p. 59.

6.   Sir William Slim, *Defeat Into Victory* (London: Cassell and Company, 1956) p. 292.

7.   Heinz Guderian, *Panzer Leader* (New York: E. P. Dutton and Co., 1952) p. 97.

8.   Ibid., pp. 105–106.

9.   Gen George S. Patton Jr., *War As I Knew It* (New York: Bantam Books, Inc., 1979) pp. 373–374.

10.   **Deception:** "Those measures designed to mislead the enemy by manipulation, distortion, or falsification of evidence to induce him to react in a manner prejudicial to his interests." (Joint Pub 1-02)

11.   Ladislas Farago, *Patton: Ordeal and Triumph* (New York: Ivan Obolensky, Inc., 1963) pp. 399–400.

12.   Slim, pp. 451–452.

13.   See Robert A. Doughty, *The Seeds of Disaster: The Development of French Army Doctrine 1919–1939* (Hamden, CT: Archon Books, 1985) p. 4.

14. B. H. Liddell Hart, *History of the Second World War* (New York: G. P. Putnam's Sons, 1970) p. 73–74.

15. The extent to which these subtleties were actually taken into account at the time is unclear. For a good examination of the problem, see Michael R. Gordon and Bernard E. Trainor, *The Generals' War: The Inside Story of the Conflict in the Gulf* (Boston: Little, Brown and Company, 1995) especially pp. 361–363.

16. Joint Pub 3-0, *Doctrine for Joint Operations* (February 1995) pp. III-9–III-10.

17. Chairman Joint Chiefs of Staff Manual 3500.04.

18. For further explanation of the importance of command and control, see chapter 1, MCDP 6, *Command and Control* (October 1996).

19. Holder, p. 123.

20. Liddell Hart, *Strategy*, p. 152.

21. Ibid., p. 339. Italics in the original.

22. **Mobility:** "A quality or capability of military forces which permits them to move from place to place while retaining the ability to fulfill their primary mission." (Joint Pub 1-02)

23. For example, the light armored vehicle has less tactical mobility than a main battle tank in most environments but has far superior operational and strategic mobility. It can be transported in

much greater numbers by strategic lift. Its comparatively simple automotive system, fuel efficiency, and wheels give it far greater operational range and speed.

24.   Patton, pp. 380–381.

25.   MCDP 1, *Warfighting* (June 1997) p. 83.

26.   **Targeting:** "2. The analysis of enemy situations relative to the commander's mission, objectives, and capabilities at the commanders' disposal, to identify and nominate specific vulnerabilities that, if expoited, will accomplish the commander's purpose through delaying, disrupting, disabling, or destroying enemy forces or resources critical to the enemy." (Joint Pub 1-02)

27.   Ronald H. Spector, *Eagle Against the Sun* (New York: Vintage Books, 1985) pp. 168–176, 448–451.

28.   Meehan, p. 16.

29.   **Logistics:** "The science of planning and carrying out the movement and maintenance of forces. In its most comprehensive sense, those aspects of military operations which deal with: a. design and development, acquisition, storage, movement, distribution, maintenance, evacuation, and disposition of materiel; b. movement, evacuation, and hospitalization of personnel; c. acquisition or construction, maintenance, operation, and disposition of facilities; and d. acquisition or furnishing of services." (Joint Pub 1-02)

30.   The distinction between strategic, operational, and tactical logistics is outlined in Joint Pub 4-0, *Doctrine for Logistic Support of Joint Operations* (January 1995) p. III-3 and MCDP 4, *Logistics* (February 1997) pp. 48–53.

31.   Maj Charles D. Melson, Evelyn A. Englander, and Capt David A. Dawson, comps., *U.S. Marines in the Persian Gulf, 1990–1991: Anthology and Annotated Bibliography* (Washington, D.C.: Headquarters, U.S. Marine Corps, History and Museums Division, 1992) pp. 158–159.

32.   Slim, p. 542

Conclusion

1.   Sun Tzu, *The Art of War*, trans. Thomas Cleary (Boston: Shambala Publications, 1988) pp. 80–81.

2.   Meehan, p. 15.

www.ingramcontent.com/pod-product-compliance
Lightning Source LLC
LaVergne TN
LVHW091155080426
835509LV00006B/699